Nuclear Energy and The Public

Social Psychology and Society
General Editors: Howard Giles and Miles Hewstone

Children and Prejudice
Frances Aboud

Contract and Conflict in Intergroup Encounters
Edited by Miles Hewstone and Rupert Brown

Interpersonal Accounts
John H. Harvey, Ann L. Weber, and Terri L. Orbuch

Nuclear Energy and The Public
Joop van der Pligt

Nuclear Energy and The Public

Joop van der Pligt

BLACKWELL
Oxford UK & Cambridge USA

Copyright © Joop van der Pligt 1992

The right of Joop van der Pligt to be identified as author of this work has been asserted in accordance with the Copyright, Designs and Patents Act 1988.

First published 1992

First published in USA 1993

Blackwell Publishers
108 Cowley Road
Oxford OX4 1JF
UK

238 Main Street, Suite 501
Cambridge, Massachusetts 02142
USA

All rights reserved. Except for the quotation of short passages for the purposes of criticism and review, no part of this publication may be reproduced, stored in a retrieval system, or transmitted, in any form or by any means, electronic, mechanical, photocopying, recording or otherwise, without the prior permission of the publisher.

Except in the United States of America, this book is sold subject to the condition that it shall not, by way of trade or otherwise, be lent, resold, hired out, or otherwise circulated without the publisher's prior consent in any form of binding or cover other than that in which it is published and without a similar condition including this condition being imposed on the subsequent purchaser.

British Library Cataloguing in Publication Data

A CIP catalogue record for this book is available from the British Library.

Library of Congress Cataloging-in-Publication Data

Pligt, J. van der (Joop)
   Nuclear energy and the public / J. van der Pligt.
     p.   cm.
   Includes bibliographical references and index.
   ISBN 0-631-18732-4 (alk. paper)
    1. Nuclear industry—Public opinion. 2. Nuclear power plants—Accidents—Public opinion. I. Title.
HD9698.A2P47 1992
333.792'4—dc20   92-16499
                                                                                                   CIP

Typeset in 10 on 12 pt Sabon by Best-set Typesetter Ltd, Hong Kong
Printed in Great Britain by T.J. Press Ltd, Padstow, Cornwall

This book is printed on acid-free paper

Contents

Series editors' preface	viii
Preface	x
Abbreviations	xiii

1 Public opinion and nuclear energy 1
   Public attitudes to nuclear energy 2
   Public opinion and the Chernobyl accident 7
   Public opinion and waste management 9
   Familiarity and public opinion 11
   Opposition to nuclear energy: two explanations 11
   Health and safety 14
   Risks and benefits of public opinion research 15
   Conclusions 16

2 Risk assessment and risk perception 18
   Risk assessment and risk analysis 18
   Risk perception 23
   Risk analysis: facts and values 34
   Conclusions 37

3 Attitudes, beliefs and values 39
   Attitudes 39
   Attitudes towards nuclear energy 41
   Dimensional salience 43
   Salience and values 44

	Perseverance of attitudes	51
	Information processing: the effects of own attitude	53
	Attitudes, consensus and person perception	54
	Conclusions	57

4 Community attitudes towards nuclear power stations — 59
- Local versus general attitudes — 59
- Local acceptance during construction — 61
- Local attitudes during siting — 63
- Salience and local attitudes — 65
- Salience and familiarity — 68
- Conclusions — 74

5 Siting nuclear waste facilities — 76
- Technological and political context — 77
- Public concern and opposition — 79
- Attitudes, risk perception and equity — 82
- Improving equity — 85
- Conclusions — 93

6 Environmental stressors — 94
- Definition of stress — 95
- Characteristics of environmental stressors — 95
- Theoretical perspectives on stress — 98
- The effects of stressors — 102
- Nuclear energy and stress — 105
- Conclusions — 110

7 Nuclear accidents: Three Mile Island and Chernobyl — 112
- Consequences of the TMI accident — 114
- Stress and psychological effects — 115
- Other effects — 119
- Conclusions about TMI — 120
- The Chernobyl accident — 120
- Public reactions to the Chernobyl accident — 124
- Institutional reactions to Chernobyl — 126
- Conclusions: the consequences of Chernobyl — 130

8 Communicating risks — 132
- Case 1: risk communication after the TMI accident — 132
- Case 2: risk communication after Chernobyl — 134

	Information and education	138
	Emergency information	140
	Policy decision-making and conflict resolution	145
	Conclusions	150
9	Decision analysis and nuclear energy policy	153
	Scenarios as decision-making tools	154
	Risk assessment and policy decision-making	159
	Multi-attribute utility theory	160
	Cost–benefit analysis	162
	Impact assessment	164
	Value-oriented social decision analysis	165
	Conclusions	167
10	Conclusions	168

Bibliography	171
Index of subjects	188
Index of names	190

Series editors' preface

On 25–26 April 1986 there was a serious accident at the Chernobyl nuclear power station in the Ukraine. It led to the largest release of radioactivity ever recorded in one technological catastrophe. From 01:23 hrs on Saturday 26 April, or rather from 21:00 hrs, Monday 28 April (when the accident was officially acknowledged by the Soviet news media), public perception of nuclear power was never to be the same again. Many European countries were faced suddenly with potential radioactive contamination, and for millions of anxious citizens the arcane language of 'half-life', 'melt-down' and 'becquerel' had obtained a frightening immediacy.

Joop van der Pligt presents a mass of compelling evidence succinctly and lucidly, illustrating convincingly the use of various social-psychological theories to reveal the public's understanding and misunderstanding of issues relating to nuclear energy. The most obvious literatures on which he draws are public opinion research from North America and Europe, analyses of risk assessment and perception, and attitude research. Less obviously, he shows how equity theory helps to explain the more negative and more extreme community attitudes towards local compared with national nuclear power stations, and reactions to the planned siting of nuclear waste facilities. It is clear that the public is concerned with who pays, either with money or health, who benefits and who enjoys what rights. The well-known NIMBY syndrome ('not in my back yard') exemplifies the desire to enjoy nuclear power's benefits, but to avoid its costs.

The approach espoused in this volume views people's reactions to nuclear issues as rational, and van der Pligt is conspicuously fair in his treatment of the various perspectives on nuclear energy. Throughout he is aware of, and compares, the different perspectives or 'frames of reference' of experts and lay people. He is also sensitive to the stresses associated with the nuclear debate and his cautious chapters on the communication of risks and decision analysis will help to raise the level of future debate in this area.

Throughout the book intelligent use is made of examples and case studies – especially the lessons to be learnt from Three Mile Island and Chernobyl. This book will, we are sure, be of great interest to readers on all sides of the nuclear debate – whether lay people or experts, pro or anti nuclear energy. In short, it could not be more true to the central aim of this series – realising the potential of social psychology to understanding and confronting contemporary social problems.

Howard Giles and Miles Hewstone

Preface

Nuclear energy has not only generated electricity but has also produced a wide variety of byproducts. Some of these were expected (for instance nuclear waste), others came as a surprise. For several decades nuclear energy has also generated a major and sustained controversy in most countries. Several major accidents and the still unresolved nuclear waste issue have led to vigorous public opposition and a virtual halt in the growth of the nuclear power industry. But the nuclear debate is likely to be reintensified again in the near future. Large-scale environmental problems, such as the heating of the atmosphere (the greenhouse effect), could well make it necessary to reconsider the costs and benefits of nuclear energy.

The importance of public reactions now seems to be accepted by all involved in the nuclear debate. This book deals with these public reactions and the factors that lie behind them. First I will present an overview of public opinion research in Europe and North America (Chapter 1). Results of this research point to the importance of safety-related issues in people's thinking about nuclear energy. Next I will focus on how people perceive nuclear risks and the characteristics of these risks that make them unacceptable to large parts of the general public (Chapter 2). Lay people's perception of risks will be contrasted with experts' risk assessments.

Chapters 3 and 4 focus on the perceived costs and benefits of nuclear energy as determinants of people's attitudes towards this energy source. First I will discuss general attitudes towards nuclear energy (Chapter 3), followed by an overview of research on local attitudes towards nuclear power stations (Chapter 4). Local attitudes during siting procedures and construction will be discussed, as well as local attitudes towards existing operational nuclear power stations.

Chapter 5 deals exclusively with siting nuclear waste facilities. It stresses the importance of equity issues. The discrepancy between possible local costs and national benefits seems to be a major source of the limited local acceptance of nuclear developments. There are several possible ways to resolve this inequity; some of these will be discussed.

Chapter 6 addresses the issues of public worry and anxiety about possible consequences of the operation of nuclear power stations; it relates these to theories of environmental stress and coping. Stress and coping also play an important role in the aftermath of serious nuclear accidents. Chapter 7 describes public reactions to the accidents at Three Mile Island and Chernobyl.

One of the factors that played a major role in public reactions to nuclear energy is communication between experts, authorities and lay people. Chapter 8 focuses on the communication of risks and presents some examples of poor risk communication after the accidents at Three Mile Island and Chernobyl. A major theme of this chapter is ways of achieving improved communications, both in a general sense and in emergency situations.

Chapter 9 focuses on the expert and discusses decision-making tools. All of these tools could help to improve the quality of policy decision-making in the context of complex issues, such as choosing between different energy sources for generating electricity. These techniques do not provide ready-made answers to the problem but could enhance the completeness and rationality of the decision-making process. Careful analysis and decomposition of the problem of possible 'energy futures' could also improve communication between the various stakeholders in the nuclear debate. Improved communication should lead to more acceptable solutions for all those involved. In all, the material included in this book should help in finding solutions. A more modest and more realistic expectation is that the present book will help to bridge the gap between the various stakeholders in the nuclear debate. Increased awareness of the frame of reference of 'the other side' could help to repair the present – often unnecessary – polarized stalemate.

I would like to thank Academic Press, The American Psychological Association, Lawrence Erlbaum, D.C. Heath & Co., The American Association for the Advancement of Science, Sage Publications, Elsevier's Scientific Publishing Company, The Society for Risk Analysis, V.H. Winston & Son Inc. and Plenum Publishing Company who gave permission to reproduce tables and/or figures included in this book.

Finally, I must express my indebtness to many colleagues whose ideas appear throughout this book. Several were active in an international research group on risk communication. They include (in alphabetical order): George Cvetkovich, Baruch Fischhoff, Roger Kasperson, Granger Morgan and Pieter-Jan Stallen. I would also like to thank people who have helped me directly with the research included in this book. Dick Eiser (University of Exeter) and Russell Spears (now at the University of Amsterdam) collaborated on some of the studies described in this book. They provided the ideal mix of (friendly) thoroughness and confidence in the applied value of psychology. Both helped tremendously in the development of my ideas about the possible contribution of (social) psychology to the nuclear energy issue. Discussions with Joop de Boer and Joop van der Linden while I worked at the Institute of Environmental Studies at the Free University in Amsterdam also helped to shape this book. I would also like to thank Miles Hewstone for his useful suggestions and Robert B. Peberdy who helped to prepare the final version of the manuscript with skill and patience. Most importantly, I thank Karin George for her unerring capability to transform unrealistic demands into possible tasks.

Joop van der Pligt
Amsterdam

Abbreviations

AIF	Atomic Industrial Forum
CEGB	(UK) Central Electricity Generating Board
DUP	Declaration d'Utilité Publique
EC	European Community
EDF	Electricité de France
FRG	Federal Republic of Germany
GAS	general adaptation syndrome
IAEA	International Atomic Energy Authority
MAFF	(UK) Ministry of Agriculture, Fisheries and Food
MAUT	multi-attribute utility theory
NIMBY	'not in my back yard'
NIREX	(UK) Nuclear Industry Radioactive Waste Executive
NRPB	(UK) National Radiological Protection Board
PAGIS	performance assessment of geological information systems
QALY	quality-adjusted life-year
SEAS	strategic environmental assessment system
THORP	thermal oxide fuel reprocessing plant
TMI	Three Mile Island (USA)
USAEC	United States Atomic Energy Commission
USERDA	United States Energy Research Development Administration
USNRC	United States Nuclear Regulatory Commission

1
Public opinion and nuclear energy

Nuclear energy has been a controversial issue since around 1980. Until the mid 1970s its acceptance as a source of electrical power seemed assured. Results of public opinion polls in the early 1970s repeatedly indicated that the majority of the public viewed nuclear energy favourably. Up to the mid 1970s the nuclear industry was convinced that the benefits of nuclear energy had been communicated to the people and that further expansion of nuclear power generation would be supported by the public. Subsequent years, however, brought more opposition to nuclear energy. It was reflected in the growth of the environmental movement, in the increasing length of public inquiries into the building of nuclear power stations, and in various local and national referenda in the United States and Europe. In the United States many initiative referenda to restrict nuclear power took place in the mid 1970s (see, e.g., Kasperson et al., 1980). Several European countries called national referenda (e.g., Austria, Sweden, Switzerland); others (e.g., the Netherlands) organized national debates about the nuclear issue. The United Kingdom saw a massive increase in the length of public inquiries into the licensing of nuclear power stations.

All these events illustrate the entrance of 'the public' into the once-exclusive domain of energy policy-making. It has become recognized that the future of nuclear energy will not only depend on technical and economic factors, but that public acceptability of nuclear energy for generating electricity will play a crucial role in energy decisions.

One consequence of these developments was that nuclear issues figured prominently in public opinion research. In this chapter I will present a brief overview of the major findings of this research. Initially the emphasis will be on US data, given the greater availability of systematic opinion research in the USA from the mid 1970s onwards. In

Europe opinion research became more prominent and coordination improved in the early 1980s (for example, with respect to the timing of surveys and the framing of questions in different languages). The second part of this chapter will thus present a wider overview of public opinion data.

Public attitudes to nuclear energy

During the 1960s hardly any public opinion surveys were conducted on nuclear energy. When Melber et al. (1977) analysed collected survey research there had been only twenty-seven national US surveys of attitudes to nuclear energy. There was one each from 1960, 1970 and 1973. There were three from 1974, eleven from 1975 and ten from 1976. This rapid increase in survey frequency illustrates the rise of nuclear energy as a controversial public issue.

This public interest is also illustrated by media coverage of the nuclear issue. Rankin and Nealey (1978) report a *fivefold* increase of US newspaper and magazine coverage between 1972 and 1976. Network television's attention to the issue increased *ninefold* between 1972 and 1977. Nealey, Melber and Rankin (1983) present an extensive overview of US public-opinion data collected between 1975 and 1981. The surveys summarized in their book show a considerable variation because of different sampling procedures, types of questions asked and means of data collection. Nonetheless, several conclusions can be drawn from the surveys conducted between 1975 and 1981. The first is that the accident at Three Mile Island (TMI) had a significant impact on public attitudes towards the construction of additional nuclear power plants. Figure 1.1 presents an overview of the public acceptability of building more nuclear power plants based on two large data sets (Cambridge Reports and Louis Harris and Associates).

The index used in figure 1.1 incorporates both support and opposition into the same metric, and is the ratio of the percentage of favourable opinions to the sum of favourable plus unfavourable opinions times 100, thus:

$$\text{Acceptability } (A) = \frac{\% \text{ in favour}}{\% \text{ in favour} + \% \text{ opposed}} \times 100$$

This index ignores the undecided response. When support and opposition are equal, A will be 50. When there is a two-to-one margin

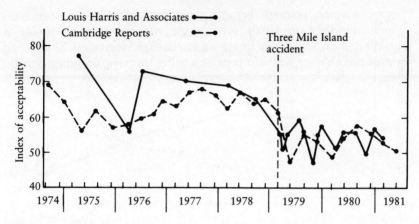

Figure 1.1 Public acceptability of building more nuclear power plants in the USA
Source: Nealey, Melber and Rankin, 1983, p. 21.

in favour A will be 67. When there is a two-to-one margin opposed to the construction of more nuclear power plants A will be 33.

As can be seen in figure 1.1, in the four years before the TMI accident the acceptability index averaged approximately 65 for both data sets. After TMI there was a substantial decline in acceptability. Although there was considerable veriation in acceptability scores in the years following the accident, acceptability did not return to its previous level. In the years after TMI, the acceptability scores were less stable and averaged just over 50 (respectively, 53.0 and 54.1). Furthermore, later US surveys showed that public attitudes towards building more nuclear power plants were substantially more negative than those reflected by the data prior to 1982 (see Nealey, Melber and Rankin, 1983). Canadian surveys of the late 1980s (Decima, 1987) showed a similar pattern with nearly 80 per cent of the total population opposing the building of more nuclear power stations. Moreover, even those in favour of the established use of nuclear energy tended to oppose expansion of the number of nuclear power stations (nearly 70 per cent). US opinion at the end of the 1980s was less extreme but still showed a majority of the public opposing the construction of more nuclear power plants. A survey carried out for *Time* magazine and CNN (Cable News Network) showed that 52 per cent of the US public opposed the construction of more nuclear power plants, 40 per cent favoured more plants and 8 per cent were unsure (*Atom*, 1991).

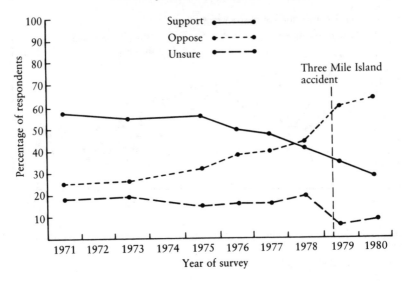

Figure 1.2 Public attitudes to local construction of a nuclear power plant in the USA
Source: Nealey, Melber and Rankin, 1983, p. 23.

Closer inspection of the pre-TMI data revealed no significant trend towards either increasing or decreasing levels of acceptability. The years immediately following TMI showed a significant increase of opposition compared with pre-TMI levels and no sign that acceptability might return to its previous level. Overall, support decreased and opposition increased. The latter increase was mainly a result of a significant decrease in uncertainty about taking a stand on the nuclear issue. Before TMI about 18 per cent of the US public was undecided; after TMI this percentage remained at about 12 per cent. The TMI accident did have a clear impact on public opinion. US polls at the time showed that more than 95 per cent of the public had heard or read about the TMI accident, 80 per cent were disturbed by it and 50 to 70 per cent believed that such an accident would happen again.

The impact of the TMI accident is also clearly reflected in opinion polls dealing with the local construction of nuclear power stations. Figure 1.2 shows the percentage of the US public expressing support for or opposition to the local construction of nuclear power plants. It shows a significant reduction of support for local construction after TMI and clearly illustrates the 'NIMBY' ('Not In My Back Yard') phenomenon. Canadian surveys confirm this picture. For instance, 58 per cent of New Brunswick's population regarded the building of a nuclear power plant in their province as a major threat; 65 per cent of Saskatchewan's residents opposed the building of a nuclear power station in their province (Decima, 1987).

The first standardized international opinion polls in the European Community were conducted in the late 1970s and early 1980s. The most elaborate survey took place in 1982 (Commission of the European Communities, 1982). Between March and May of 1982 an identical set of twenty questions was presented to national representative samples of the population aged fifteen and over in the ten countries of the European Community (total $N = 9,700$). Table 1.1 presents an overview of opinion shifts between 1978 and 1982 and shows considerable variations between the ten EC member states.

France (the EC country that had the most nuclear power stations) was the country with the most favourable public opinion. In 1982 a small majority of the French population thought the development of nuclear power worthwhile. German public opinion also moved in

Table 1.1 Attitudes in EC countries towards further development of nuclear power 1978–82

Country	Development is worthwhile			Development leads to unacceptable risks		
	1978 %	1982 %	Difference	1978 %	1982 %	Difference
France	40	51	+11	42	31	−11
Federal Republic of Germany	35	37	+2	45	30	−15
Netherlands	28	34	+6	54	48	−6
Belgium	29	27	−2	39	37	−2
Luxembourg	35	32	−3	31	49	+18
Denmark	37	25	−12	34	49	+15
United Kingdom	57	39	−18	25	37	+12
Italy	53	34	−19	29	43	+14
Ireland	43	13	−30	35	47	+12
Greece	*	15	–	*	50	–
Overall mean		38			37	

* Question not asked.

Overall means are corrected for sample size differences.

Source: Adapted from Commission of the European Communities (1982), p. 38.

Table 1.2 Opinions in EC countries about further development of nuclear power as a function of the scale on which nuclear power is already developed

Country	Worth while %	No particular interest %	Unacceptable risks %	Don't know %
Large-scale nuclear development				
Belgium	27	9	37	27
Federal Republic of Germany	37	14	27	22
France	51	4	30	15
United Kingdom	39	17	37	7
Moderate nuclear development				
Italy	34	5	42	19
Netherlands	34	6	48	12
No nuclear power				
Denmark	25	9	50	16
Greece	15	6	49	30
Ireland	13	21	47	19
Luxembourg	32	8	49	11

Source: Adapted from Commission of the European Communities (1982), p. 35.

favour of higher acceptability but there was no majority of supporters because of the large percentage of undecided people. This was reflected in local and federal opposition to expansion of the nuclear industry. As can be seen from table 1.1, several countries saw a steady increase of opposition, mainly at the expense of the undecided category. The overall percentage of support remained at about 37.

Table 1.2 gives a more discriminating overview of the 1982 survey. In this table countries are grouped in terms of the scale on which they had developed nuclear energy. The first group consists of countries in which nuclear energy had been developed on a large scale (France, FRG, UK and Belgium); the second group of countries that had a relatively modest number of nuclear power plants (the Netherlands and Italy); the third group of countries that had no nuclear power stations (Luxembourg, Denmark, Ireland and Greece).

The basic question presented to all respondents concerned the desirability of developing nuclear power plants to generate electricity. People were given four alternative responses: (1) it is worthwhile; (2) no particular interest; (3) involves unacceptable risks; (4) don't know. Overall percentages in the four categories (after adjusting for sample size differences) were, respectively, 38 per cent, 10 per cent, 37 per cent and 15 per cent.

It is clear that the public was divided about the issue and that over 25 per cent of European citizens had either no opinion or found the issue not particularly interesting. At a national level, public opinion was more favourable in countries in which nuclear power had been developed on a larger scale. The risks are seen as most unacceptable in those countries that had no nuclear power stations or only a small number.

Further breakdowns of these figures according to political preference, age and educational level revealed the following patterns. First, the political left–right dimension was clearly correlated with views about nuclear development. A majority (52 per cent) of those on the right of this political dimension thought the development was worthwhile, compared with just under 30 per cent for those on the left and 40 per cent for centre positions. Of those on the left, between 47 and 60 per cent saw further nuclear development as an unacceptable risk. The percentage for those on the right was only about 25 per cent.

Much weaker relationships were found with age (with younger people being more opposed) and educational level. Level of education was not significantly related to the perceived acceptability of the risks of nuclear developments. The main effects of educational level were: (a) much higher percentages of 'don't knows' and 'not interested' for the lower education groups (32 per cent versus 13 per cent); (b) higher levels of support from the higher educational groups; i.e., these groups saw development as worthwhile (49 per cent versus 34 per cent).

Public opinion and the Chernobyl accident

One of the issues discussed in the previous section was the impact of major accidents on public opinion about nuclear energy. The impact of the TMI accident is clearly illustrated by US data showing a significant increase in public opposition and no clear recovery of public support to pre-TMI levels.

Seven years after the TMI accident, in 1986, a much more serious accident occurred at Chernobyl in Ukraine (then a republic within the USSR). It caused radioactive contamination in many European countries. Not surprisingly extensive opinion surveys were carried out after the accident, both in Europe and in the USA.

The Chernobyl accident produced similar effects on public opinion as the TMI accident had done. Support for nuclear energy declined in most

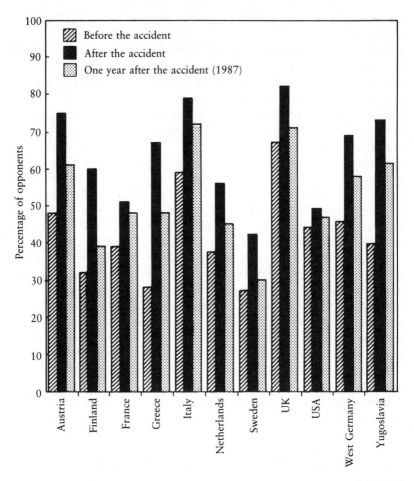

Figure 1.3 The Chernobyl accident and public opinion in Europe and the USA
Source: Renn, 1990, p. 156.

countries. Some later recovery was noticeable, but generally support did not return to pre-Chernobyl levels. Immediate increases in public opposition were very large in Finland, Yugoslavia and Greece (over 30 per cent), substantial in Austria, FRG and Italy (over 20 per cent) while moderate changes took place in the UK, the Netherlands, France and Sweden (12–18 per cent). Figure 1.3 gives an overview of these changes (see also Suhonen and Virtanen, 1987; Renn, 1990).

Although the USA was not affected by the fallout from Chernobyl, public opposition to nuclear energy increased by 5 per cent to reach a peak of 49 per cent, the highest ever reported (*Newsweek*, 1986). More dramatic changes can be seen in public attitudes towards building new nuclear power stations in people's localities. Opposition to the building of a 'local' nuclear power station was already quite substantial after the TMI accident. After Chernobyl it increased further, to 70 per cent. However, within a year the level of opposition in the USA returned to pre-Chernobyl levels (Renn, 1990). Elsewhere there were different reaction patterns. In Canada, over a year after Chernobyl, 79 per cent of the population disagreed with the statement that 'there is really no chance that there will be a major accident in Canada's nuclear power reactors like the accident at Chernobyl' (Decima, 1987). Moreover, respondents tended to think that the most likely cause of such an accident would be with human error (54 per cent), followed by operators taking short-cuts (33 per cent).

Figure 1.3 shows the increases of opposition as measured within two to three months after the Chernobyl accident and an indication of the stability of these changes (survey data collected at least a year after the accident). Most countries show a reduction of public opposition a year after the accident. It needs to be added, however, that this return to pre-accident levels was not complete. A year after the accident public opposition was still 5 to 25 per cent higher than before the accident.

Although no public opinion data are available from East European countries, observers reported a growing opposition to nuclear energy in Poland, Hungary and Czechoslovakia (see, for example, *Nucleonic Week*, 1986; Renn, 1990). Environmental issues also played a major role in the dramatic changes in Eastern Europe that began in 1989.

One longer term effect of Chernobyl was that it influenced growing opposition to the siting of nuclear waste facilities. To this issue I now turn.

Public opinion and waste management

Public concern about nuclear-waste management increased steadily from 1975 and through the 1980s. Nealey, Melber and Rankin (1983) and Kasperson et al. (1980) conclude that the public sees nuclear-waste management as one of the most important, if not the most important, issue of nuclear energy. US data reported by Melber et al. (1977) indicate that before 1977 the public was more or less evenly split about whether satisfactory waste-disposal options were available. About 40 per cent believed that safe waste-disposal solutions were available; 40

per cent did not; 20 per cent were uncertain. About 50 per cent thought that safe waste-disposal methods would be developed; 25 per cent did not. Later surveys reveal a less optimistic view about technical solutions to the waste-disposal problem.

Attitudes towards the storage of nuclear wastes in different types of 'hosts' (granite, rock salt, seabed, above-surface structures in isolated areas) revealed most acceptance for above-surface storage (53 per cent), followed by storage in granite (44 per cent) and rock salt (43 per cent). Seabed disposal was seen as the least attractive solution (30 per cent). Nealey, Melber and Rankin (1983) also report data illustrating the NIMBY ('Not In My Back Yard') phenomenon. More than 30 per cent of the general public said they would be more likely to support waste disposal if the site selected for permanent disposal was not in their state. Politicians were most influenced by this condition: 63 per cent would be more likely to support disposal if it was not done in their own state.

Another finding reported by Nealey et al. is that since 1976 over twice as many people have said that they oppose the further expansion of nuclear energy because of unresolved waste-management issues than was previously the case. Moreover, when respondents who had changed their attitude towards nuclear energy from support to opposition were asked why, the reason mentioned most often was increased concern over nuclear-waste management. This concern was enhanced by the TMI-accident, which suggested that any serious nuclear accident would increase the salience of the nuclear-waste issue. Finally, when people were asked to make a direct comparison between waste management and reactor safety, respondents believed the former to be a bigger problem (Nealey, Melber and Rankin, 1983). Results obtained in Canada show a similar pattern. A total of 56 per cent of respondents indicated that nuclear-waste disposal was of greater concern than the potential of an accident at a nuclear reactor (41 per cent). More than 60 per cent of the respondents thought the nuclear industry to be incapable of handling waste disposal. Not surprisingly, nearly 70 per cent of local residents opposed a waste facility in Northern Ontario (Decima, 1987).

European surveys found similar attitudes. The 1982 survey in 10 EC member states presented respondents with several possible hazards associated with nuclear power. They were asked to select the hazards that worried them most. A total of 57 per cent selected the nuclear-waste issue; 51 per cent routine (radioactive) emissions from nuclear power stations; 23 per cent selected a major accident (explosion) of a nuclear power station. Later polls conducted in the UK showed a similar pattern. When respondents were confronted with a variety of environmental hazards, ranging from 'lead in petrol' to domestic waste and air pollution, they selected nuclear waste dumping as the number one issue – 64 per cent of the respondents were worried about this issue (*Atom*, 1990). The nuclear-waste issue thus figures prominently in the general public's beliefs about nuclear energy. Chapter 5 will discuss this issue in more detail.

Familiarity and public opinion

Several surveys have compared the level of acceptance of nuclear energy among people who live near a power station with that of people who do not. Other surveys have monitored public opinion in a locality where the possibility of a nuclear power facility being constructed gradually becomes a reality. Overall, there is mixed support for the idea that familiarity leads to greater acceptance of a nuclear facility.

Melber et al. (1977) mention eight studies which measured local acceptance of a nuclear power plant as it was being constructed. Only two of these showed a significant increase in acceptance over time. In one locality there was a significant increase in the level of opposition. Hughey et al. (1983) found large negative changes in attitude towards a nuclear facility while it was being constructed. Their study indicated that people had much lower expectations about potential positive outcomes in the later stages of construction. Results are equally mixed concerning the relationship between living near a nuclear power plant and acceptance of nuclear energy in general (see Thomas and Baillie, 1982). A study by van der Pligt, Eiser and Spears (1986b) showed marginally more favourable attitudes towards nuclear energy in general around Hinkley Point (the site of two existing nuclear power stations in the Southwest England) than in three small local communities that were selected as possible future sites for stations. The 1982 EC survey also suggested greater acceptance in countries which had a higher number of nuclear power stations. Within countries, however, there was little difference in attitude between those who lived within 48 to 96 km of a power stations and other respondents. Other research on local attitudes does not support the notion that familiarity leads to more favourable attitudes (Warren, 1981).

Opposition to nuclear energy: two explanations

This section discusses the two main explanations which have been put forward for the reported increases in public opposition. One emphasizes the emotional, irrational aspects of human reactions to nuclear issues. For instance, Lifton (1979) argues that the imagery of nuclear weapons has had a profound effect on the human subconscious, and that people's fear of nuclear energy is an extension of their fear of nuclear weapons. Dupont (1981) characterizes the fear as phobic and believes that this 'ultimate irrational fear' is enhanced by the media's focus on fear in their coverage of the nuclear debate.

Hohenemser, Kasperson, and Kates (1977) also put forward the view that the distrust of nuclear power is rooted in the (American) fear of the bomb. There are some arguments against this view. As was noted by Nealey, Melber and Rankin (1983), when the first demonstration nuclear power plant to produce electricity for public energy consumption went on line in 1958 at Shippingport, Pennsylvania, few voices were raised in protest. Moreover, President Dwight D. Eisenhower's 'Atoms for Peace' programme launched in 1956 was a response to the growing controversy over nuclear testing and was presented as a welcome alternative, stressing the potential for non-destructive uses of nuclear energy.

This optimism was still apparent in 1960 when 64 per cent of the US public was in favour of nuclear power as a source for electricity (6 per cent was against and 30 per cent had no opinion or was undecided). It was not until the mid 1970s that things began to change. Of course, this does not imply that the issues of nuclear energy and nuclear weapons are not related. In fact in quite a few countries nuclear power generation and nuclear weapons are related to the extent that plutonium, a byproduct from some types of nuclear power station, is used for the production of nuclear weapons. Moreover, surveys indicate a link between attitudes to the two issues, especially in countries where both technologies are used. Another similarity between these two issues is that both (nuclear) energy policy and (nuclear) defense policy tended to remain in the domain of elite decision-making. Both issues then moved into the public eye. In both areas public opinion polarized, consensus broke down and the public entered once exclusive domains of policy decision-making (see March and Fraser, 1989). This does not imply, however, that 'irrational fear' provides the best explanation for public reactions to nuclear power.

The second explanation of increasing public opposition generally assumes that rational processes underlie people's decisions about the acceptance of nuclear facilities. In this tradition, public concern about nuclear facilities is based on integrating information and relating it to one's own (subjective) values in order to reach an overall judgement. In other words, public opposition is not caused by subliminal fears or phobias but is based on everyday inferential strategies.

Explanations that stress the importance of emotional, irrational aspects in opinion formation are usually favoured by those supporting nuclear power, with the restriction that these subliminal drives are seen as mainly applicable to those opposing nuclear energy. It seems unlikely, however, that the anti-nuclear side of the debate should have a monopoly on subconscious motivations. Furthermore, both sides of the debate rely upon factual arguments and both use legal and scientific experts to strengthen their case. These arguments refer to a wide variety of aspects (economic, political, environmental). Strictly speaking, all these rational arguments could still be traced to subconscious elements, making it difficult to test the relative validity of the two approaches.

Mitchell (1984) attempted to compare the two approaches and concluded that the 'lay rationality' approach had greater explanatory power. Granberg and Halmberg (1986) also looked at the rationality of the public, in this case the voting behaviour of the Swedish population in the national referendum on nuclear power of March 1980. Their findings provided most support for what they termed a 'rational democratic' model which assumed voters to be relatively well-informed, autonomous and rational decision-makers. Finally, findings concerning the relationship between knowledge, familiarity and attitudes are mixed (see the previous section) and do not provide consistent support for the view that nuclear opponents are less knowledgeable about nuclear issues.

The major advantage of approaches that assume a rational public is that they provide an explanatory framework that can incorporate a variety of aspects seen as relevant. Approaches emphasizing the emotional, uninformed character of public reactions tend to under-estimate the relevance of economic, political and environmental arguments by reducing them to manifestations of irrational fear. Although 'lay rationality' approaches have tended to underestimate the role of fear, developments after the mid 1980s focused on decision-making models that incorporated more motivational processes, such as reactions to stress and coping mechanisms.

In the remainder of this book, therefore, I will focus on approaches that view people's reactions to the nuclear issue as rational. Within this approach I have attempted to relate people's fears to judgmental and inferential processes that determine people's evaluation of nuclear risks. The factor that has been mentioned most often in public opinion research as the major reason for public concern is the safety issue. This possible explanation for public opposition will be discussed next.

Health and safety

Only a limited amount of survey data is available that deals directly with questions of health and safety. Nealy, Melber and Rankin (1983) report several surveys on nuclear energy in which respondents were asked to volunteer the reason for holding their attitude. Their responses were classified into six categories: danger responses (accidents, nuclear-waste issue, health-related responses); no need for more plants; economic costs; pollution; insufficient technical knowledge; and the belief that other alternatives are better. Findings showed that between 50 and 75 per cent of those who opposed nuclear energy gave danger-related reasons for holding their attitude.

Other surveys revealed similar findings. The harmful consequences mentioned most often concerned reactor accidents, leakage of low-level radiation, nuclear-waste management and effects on the health of present and future generations (Nealey, Melber and Rankin, 1983). In general, US surveys indicate that the public sees nuclear power as an energy source with no shortage of fuel (wanium) which is also regarded as non-polluting in terms of gaseous emissions. It is also, however, considered the least-safe energy source and one of the most expensive. Canadian surveys reveal a similar concern about safety issues, with 42 per cent of the population indicating a willingness to live with the risk of an accident (if contained within the plant) and 43 per cent stating that they are simply unwilling to live with any risk of a nuclear accident (Decima, 1987). This survey also indicated that spontaneous responses to nuclear power tended to be dominated by safety issues.

European surveys also point to the importance of the safety issue. Respondents in the 1982 EC survey were asked to select the three most risky technologies from a list to ten. The first three were a chemical factory (selected by 71 per cent), an ammunition/explosives factory (64 per cent) and a nuclear power station (60 per cent). The 1982 survey also attempted to investigate the public's image of nuclear power stations. Results indicated consensus about the view that nuclear power can provide additional electricity and that its development could have favourable economic effects. A majority of the respondents agreed with these views. Statements expressing the view that nuclear energy is a clean form of energy and that it is cheaper to produce resulted in minority support and substantial proportions of the public not knowing whether to agree or disagree with each of these views. For instance, nearly 40 per cent did not know whether to agree or disagree with the statement that nuclear energy is cheaper than other energy sources.

The safety issue clearly divided the public. About 40 per cent agreed with the assertion that 'safety measures are so strict that they eliminate nearly all the danger', while 37 per cent disagreed. Furthermore, there was considerable consensus on a number of safety issues. Firstly, 73 per cent of the total sample (both opponents and proponents) agreed with the statement that 'nuclear power stations can be dangerous for the people who work in them'.

Nearly 70 per cent agreed with the statements 'Numerous safety precautions clearly indicate that by their very nature nuclear power stations are dangerous' and 'The increase in the number of nuclear power stations is dangerous'. Safety issues tend to be even more important to those opposing nuclear energy. Over 90 per cent of opponents agreed with the three statements concerning the possible dangers of nuclear energy. Findings obtained in a longitudinal study in the Netherlands (Verplanken, 1989) also underline the predominance of safety issues in the public's perception and evaluation of the nuclear issue.

The central role of these safety issues has resulted in a substantial research effort into public responses to technological risks in general and to risks from energy systems in particular. The next chapter will focus on the issue of risk.

Risks and benefits of public opinion research

This chapter has presented an overview of some of the major, large-scale surveys conducted in the USA, Canada and Europe. The major benefit of these surveys is their longitudinal character, enabling us to look at changes over time. Surveys which over time pose identical questions and offer the same set of response alternatives allow certain comparative conclusions. However, a few cautionary notes are in order.

First, although the surveys discussed here are longitudinal they are not based on the same respondents. Hence it is difficult to study changes over time within a specific group. The conclusion that attitudes are stable could therefore be premature. For instance, Midden and Verplanken (1990) found that aggregated stability over time does not necessarily preclude substantial attitudinal shifts that cancel each other out.

Second, one should be extremely careful in comparing opinion poll results without considering the exact wording of the question and response alternatives. Questions aimed at tapping people's opinion about an issue can be framed in a variety of ways. This is also true in the area of nuclear energy. Thomas and Baillie (1982) found at least six different types of questions which they suggest could tap different aspects of the issue at hand. They further pointed out that the meaningfulness of answers will, to a large extent, depend on the way the questions are framed. Roisser (1983) makes a similar point about possible biases due to question framing. Van der Pligt, Eiser, and Spears (1987a) point to the role of 'availability'. Their subjects were asked to answer a series of questions about nuclear energy and related issues. Their findings indicated that just being *asked* about various alternative sources of energy is enough to make them more salient and to be considered more seriously. Their study serves as a reminder of the substantial effects of context variables on opinion and attitude surveys. For all these reasons one should be careful about making cross-cultural and historical comparisons if the questions and/or response categories are not comparable. The various surveys discussed in this chapter allow comparisons over time on the basis of US and EC survey results. Direct comparisons between the outcomes obtained in the two continents are, however, more difficult to make. Finally, opinions are usually based on more specific beliefs about risks and benefits. Most surveys are limited to a few questions, making it difficult to study both structure and process in more detail. This will be attempted in the remaining chapters of this book.

Conclusions

This chapter has presented a brief overview of public opinion research with regard to nuclear power. Results of public opinion surveys show that general attitudes towards nuclear power plant construction remained relatively stable in the mid 1970s. After the accident at Three Mile Island in 1979, US attitudes became less pro-nuclear and more anti-nuclear. Moreover public support for nuclear power did not return to pre-TMI levels. The Chernobyl accident in 1986 had a similar impact on public opinion. This impact was dramatic in some European countries and also resulted in more anti-nuclear opinions that remained relatively stable. Thus, serious nuclear accidents have shown the ability to influence beliefs and attitudes towards nuclear power.

In 1983 Nealey, Melber and Rankin concluded that attitudes will be likely to change in the future if significant nuclear-related events occur.

They further predicted that future reactor accidents could have a further negative impact on public opposition to nuclear power, especially if any accidents were to occur within several years of TMI. The data on the Chernobyl accident confirm their prediction. One remark is in order, however. It seems that the effects of such accidents are less long-lasting when the geographical distance between an accident and the public is large and when the consequences are limited. For instance, a year after Chernobyl US public support more or less returned to pre-Chernobyl levels. This was not the case in Europe. Similarly, the return to pre-TMI levels was probably more marked in Europe than in the USA (see, for example, Midden and Verplanken, 1990). Overall, public opposition and support for nuclear power seem to be evenly balanced in both the USA and Europe.

Support for local plant construction decreased gradually and significantly through the 1970s, while opposition increased. Local attitudes seem now firmly against the local siting of nuclear power plants both in Europe and the USA (see, for example, Nealey, Melber and Rankin, 1983; van der Pligt, Eiser and Spears, 1986a, 1986b, 1987b). Most research findings indicate a majority of people being opposed to the building of a new nuclear power station in their locality.

One of the major obstacles to public acceptance of nuclear energy is safety. The survey data presented in this chapter indicate that the public's reactions are dominated by the risk of major accidents and the negative consequences of routine emissions from nuclear power stations and/or waste facilities. Proponents of nuclear energy also mention these aspects but seem more optimistic about our ability to prevent them.

2

Risk assessment and risk perception

Since the 1950s the rapid development of chemical and nuclear technologies has been accompanied by the potential to cause catastrophic and long-lasting damage to both the environment and public health. The processes underlying these complex technologies are unfamiliar and generally incomprehensible to most lay people. Quite often the most harmful consequences are rare and delayed, are difficult to assess by statistical analysis and are certainly not well suited to management by trial-and-error learning (cf. Slovic, 1987). These developments led to the creation of a new technique called risk assessment, designed to help identify and quantify risk.

As we have seen in the previous chapter, safety issues play an important role in public acceptance of nuclear energy. In this chapter we will first discuss risk assessment techniques, tools used by experts to decide about the acceptability of technological hazards. Next we will discuss research in the area of risk perception and describe differences between lay people and experts in their assessment and acceptance of risks. Finally, we will place this research in the wider context of the nuclear debate and evaluate the contribution of research on risk perception to our understanding of public reactions to this technology.

Risk assessment and risk analysis

Policy decision-making has become increasingly dependent on quantitative risk assessment. Generally risk assessment proceeds in four steps: (1) hazard assessment; (2) dose-response assessment; (3) exposure assessment; (4) risk characterization (see Russell and Gruber, 1987).

Hazard assessment examines the evidence that relates exposure to a specific agent with its toxicity and should, if possible, result in a qualitative judgement about the strength of that evidence. In this first stage evidence from both human epidemiology and the laboratory testing of animals is important. This stage thus focuses on the possible adverse consequences for health and the environment of specific toxic agents.

Next, *dose-response assessment* investigates the quantitative relation between specific exposure levels of a toxicant and the incidence or severity of a response in organisms. Research on dose-response relations tends to rely on test animals; on the basis of these results one draws inferences for humans. Occasionally, dose-response assessment can be studied in human samples. This is usually as a result of accidents in which groups of people (e.g., workers, local population) are exposed to a toxicant. Generally, knowledge about dose-response relations is very limited. For instance, there are over 100,000 chemical compounds with possible adverse consequences for the environment and public health. Research on quantitative dose-response relations is bound to be a very elaborate process. Even in the field of research most relevant to our purposes (radioactive toxicants) there is only limited knowledge about the dose-response relationship of long-term effects of low-level radiation and possible effects to unborn life. Not surprisingly, the 'cocktail' effects of exposure to several toxicants are even less well understood.

Exposure assessment identifies populations which could be exposed to a toxicant, describes their composition and size and also investigates the possible routes, magnitudes, frequencies and durations of such exposure.

Finally, *risk characterization* presents a synopsis of all the available information and should help policy-makers to reach conclusions about the nature of risks. This step also includes an evaluation of the magnitude of the uncertainties involved, and of the major assumptions that were used (Russell and Gruber, 1987). This synopsis should then form the basis of regulations and risk management.

The last element in risk assessment – risk characterization – is crucial in understanding what risk assessment can and cannot do. Any risk assessment contains decision points where risk to human health can only be inferred. Unfortunately, there are also a number of aspects which tend to be difficult to incorporate in quantitative risk analysis. For instance, unplanned events that might occur during normal operations of a nuclear power station are not easily analysed and quantified. In recent years these unplanned events have received considerable attention, mainly because of their role in some of the major accidents in the nuclear industry. Identification of possible unplanned events that might occur during operation (e.g., accidents, sabotage or misuse), identification of their consequences and analyses of the corresponding probability distribution have all been extensively studied (see, for instance, USNRC, 1975). Progress has been made but, as argued before, factors like human error are very difficult to quantify.

These high levels of uncertainty in risk assessment techniques have contributed to the controversies that surround the use of risk assessment as a guide to making regulations. Another major contributor to this controversy (as noted by Russell and Gruber, 1987) may arise from the confusion between risk assessment, a largely scientific enterprise that focuses on technical considerations, and risk management. The latter refers to the process by which a regulatory agency decides what to do about the results of a risk assessment. A risk management decision may involve economic, social and political considerations. Too often, risk management activities have been based on a combination of economic and technical considerations, resulting in a limited separation between risk assessment and risk management.

An important consequence of the above developments was that large-scale risk assessment studies such as the *Reactor Safety Study* of the US Nuclear Regulatory Commission remained controversial (USNRC, 1975). For several years after the publication of the *Reactor Safety Study* probabilistic risk assessment remained primarily a research topic little used by the regulatory authorities (Russell and Gruber, 1987). After the accident at Three Mile Island, in 1979, the USNRC's Advisory Committee on Reactor Safeguards recommended that more consideration should be given to the establishment of quantitative safety goals for nuclear power reactors. A number of other relevant bodies also recommended that safety objectives and the underlying philosophy be better articulated and thoroughly communicated to the public (Okrent, 1987). Similar conclusions were drawn in Canada. Decima (1987) found that Canadians clearly believed that the nuclear industry did not communicate with the general public, was not open or accountable and did not provide information.

In 1986 the USNRC published a 'Policy Statement' on safety goals. This stated the following *qualitative* safety goals: (a) individual members of the public should be provided with a level of protection from the consequences of nuclear power plant operations such that individuals bear no significant additional risk to life and health; (b) the risk to society's life and health from nuclear power plant operation should be comparable to or less than the risk from generating electricity by viable competing technologies and should not be a significant addition to other societal risks (USNRC, 1986). It was further determined that the chance of a very large release of radioactive materials to the environment should be less than 10^{-6} per reactor year. This guideline was more conservative than previous safety goals. For instance, when the Atomic Industrial Forum (AIF, 1981) proposed a framework for establishing and using quantitative safety goals it suggested a much higher goal level. In the early 1980s regulatory agencies accepted goal levels of 10^{-6} and 10^{-5} per site per year for early and latent cancer mortality risks. Later these were reformulated as not to exceed 0.1 per cent of the normal background accident or cancer mortality risk (USNRC, 1986).

Other developments in risk analysis and risk assessment attempted to look beyond expected fatalities (for which risk assessment of nuclear energy was criticised). Renn and Swaton (1984) argued that the formal definition of risk (magnitude of losses x probability of occurrence) stemmed from a narrow viewpoint. They proposed a number of steps to widen the concept of risk. First, consensus about what is regarded as harmful or beneficial. Secondly, agreement on how to combine different adverse consequences into one overall measure. Thirdly, increased attention to secondary effects. It became accepted that using the number of immediate fatalities as a total measure of the damage to society is insufficient. Risk assessment measurements should therefore also take account of other consequences. Tasker (1989b), on the basis of the work of C.F. Clement (of the Harwell Laboratory, UK), suggested that societal risk analysis should incorporate both personal and financial consequences. Personal risks are those affecting people directly. In order to categorize these, Clement suggested constructing a risk profile of possible consequences: (1) early death; (2) delayed death; (3) serious incapacity; (4) forced (permanent) evacuation; (5) serious birth defects. The societal risk for each category (for example, in case of a nuclear accident) can be obtained by a combination of counting the number of individuals affected in each category and estimating the delayed effects.

The resulting overall personal-societal risk can then be summed over the five categories using weighting factors to compare the risk in each category with that of the worst risk category (early death). This results in the following equation (2.1):

$$N = \sum_{i=1}^{5} n_i a_i$$

where i represents each of the five risk profile categories, n_i the number of cases in each category and a_i their weighting factor compared to that of early death ($a_1 = 1$). This more elaborate type of risk analysis has, apart from the problem of uncertainty (described earlier), another complicated characteristic: assigning weights to the various categories. Some progress has been made in this area (e.g., comparison of early and delayed death could be made on the basis of loss of life expectancy), but many problems remain. Tasker (1989b) suggested the application of the concept of the quality-adjusted life-year (QALY), which is used as a tool to study value for money in health care. This calculation entails multiplying each remaining year of life by a specific fraction in order to weigh the impairment in its quality experienced by the survivors. Applications of this concept are relatively new; more research is needed to find generally acceptable solutions to weigh the possible adverse consequences of hazards such as nuclear energy. Concepts such as QALY could help to find acceptable solutions to the necessary development of risk assessment techniques which include a wider range of consequences than the expected number of immediate fatalities.

Societal risks not only include the personal consequences discussed above, but also possible financial losses for society as a whole. For instance, the nuclear accident at Three Mile Island scored relatively low on the five personal risk categories (much lower than, for example, the chemical plant accident at Bhopal in India in 1984). Nonetheless, it was a major accident. This becomes clear when one looks at the financial losses: the costs of dismantling and cleaning up the plant at TMI are put at around $1 billion. The Chernobyl accident scored higher on both counts. Early deaths were about 30; the estimated number of delayed deaths from cancer is about 15,000 (with a large uncertainty range). Financial losses are bound to be enormous. In Western Europe alone the lost revenues from agriculture amounted to around $100 million.

These examples illustrate the aim of this more elaborate form of societal risk analysis: to link together the personal risks and the financial costs to society in order to quantify the total cost (T) of (possible) accidents or disasters. This can be summarized by the equation (2.2):

$$T = C + bN$$

where T refers to the total costs, C to the financial losses, and b represents the value assigned to a human life, multiplied by N (overall personal risks as calculated by equation 2.1).

Attempts like these aim to quantify risk to society and to express in monetary terms the overt costs C to society of an accident or disaster (loss of output, cost of clean-up, lost revenues) and the collective personal risks (in terms of expected loss of life). Although other potentially important factors are not yet included (e.g., damage to the environment, less dramatic long-term health effects) it seems a more complete approach than early attempts which focused entirely upon the loss of life. Values for the weighting factors of N and b (the value of a human life) bring the method into the realm of risk management and policy-making. Moreover, these weighting factors also stress the need to fill the gap between risk assessment and risk management. Both the measurement of risk and the weighting of personal and financial risks should help to make possible rational choices about alternative courses of actions.

Overall, quantitative safety goals have become increasingly important in the regulatory process. Politicians, regulatory agencies and experts frequently refer to these probabilistic criteria and employ risk assessment to evaluate hazards. Although the scientific basis for risk assessment is often uncertain and the public and its representatives have often been confused by its use in policy decisions, there is a growing dependence on quantitative risk assessment. In this context it is crucial to improve public understanding of the meaning of risk assessment. The majority of people, however, rely on intuitive risk judgements. There is a substantial body of research that studied the way in which lay people perceive and assess risks. Successful communication between experts and lay people requires a better understanding of ordinary conceptions of risk. Research on this issue is described in the next section.

Risk perception

The concept of risk and the notion of uncertainty are closely related. We may say that the lifetime risk of a stroke is 12 per cent, meaning that approximately 12 per cent of all people will suffer a stroke in their lifetime. Psychological research on risk perception originated in empirical studies of probability assessment and decision-making processes (Edwards, 1954). A major development in this area has been the discovery of a series of cognitive strategies, or heuristics, that people use when dealing with uncertainty. Generally, people have difficulties in understanding probabilistic processes. Sometimes uncertainty is simply denied, sometimes misjudged and often one is overconfident about one's judgement. Interestingly, experts appear to be prone to many of the same biases as those of lay people, especially when experts cannot rely upon solid data and have to use their intuition (see, e.g., Kahneman, Slovic, and Tversky, 1982).

Initially, research on risk perception attempted to develop a taxonomy for hazards in order to understand and predict public responses to risks.

In this way it was attempted to improve our insight into why some hazards led to extreme aversion and others to indifference. Furthermore, it was hoped that the findings would also help to explain the discrepancies between public reactions and the opinions of experts. The early efforts of Starr (1969) had a clear impact upon the type of questions this field of research addressed. Starr attempted to develop a method for weighting technological risks against benefits in order to answer the question 'How safe is safe enough?'. One of his major conclusions was that the public is much more likely to accept risks from *voluntary* activities than from *involuntary* activities. Voluntary risks up to 1,000 times greater than involuntary risks with the same level of benefits were seen as equally acceptable.

The emphasis in Starr's study (1969) was to develop a taxonomy in order to explain differences in the acceptability of risks as a function of specific characteristics of these risks. This line of research assumes that controversies about technological developments are conflicts about *risk perception* and the acceptability of risk (risk being defined as human fatalities per unit of exposure), *economic benefits* (measured in dollars) and the dichotomy *'voluntary'* versus *'involuntary'*. This classification should explain revealed patterns of risk acceptance. In a later study Otway and Cohen (1975) attempted to replicate the risk–benefit relationship obtained by Starr but their findings did not provide clear support for Starr's findings. Lowrance (1976) classified risks using a total of 10 dimensions. His aim was to clarify the nature of conflicts about technological risks and improve the understanding of public acceptability of these risks. Table 2.1 describes the dimensions used by Lowrance.

Generally, results indicated substantial differences in the acceptable levels of risk as a function of risk characteristics. Factors such as

Table 2.1 Risk dimensions used by Lowrance (1976)

---

voluntary – involuntary
immediate effect – delayed effect
alternatives available – no alternatives available
risks known with certainty – risks not known
exposure is an essential – exposure is a luxury
occupational – non-occupational
common hazard – dread hazard
affects average people – affects sensitive people
will be used as intended – likely to be misused
reversible consequences – irreversible consequences

---

voluntariness and familiarity were clearly related to acceptance; voluntary and familiar risks generally stand a better chance of being seen as acceptable.

Slovic, Fischhoff, Lichtenstein and colleagues conducted a series of studies of expressed preferences. Their aim was to show that perceived risk is both predictable and quantifiable. In one of these studies, subjects were asked to give their subjective estimates of the frequency of death from 41 sources (Lichtenstein et al., 1978). Results show that people were approximately accurate but that their judgements were systematically distorted.

Overall, their data suggested that people have a consistent subjective scale of frequency that is relatively unaffected by different ways of asking them to assess the various risks. Furthermore, these judgements

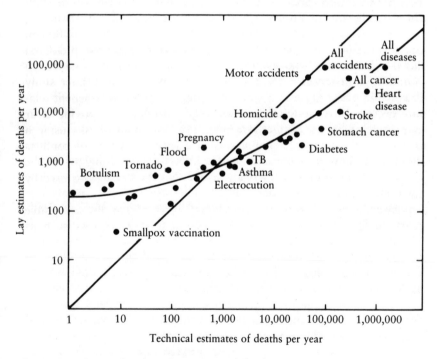

Figure 2.1 Lay and technical estimates of deaths
Source: Lichtenstein et al., 1978, p. 565.
Note: If lay estimates and technical estimates were equal, the data would fall on the straight line. The points, and the curved line fitted to them, represent the average responses of lay people. Of the activities presented to subjects only a selection has been labelled.

correlated fairly well with available statistical estimates. Figure 2.1 summarizes their findings. If judged and actual frequencies were equal, the data would fall on the straight line. The points, and the fitted curved line, represent the average responses of the sample of lay people. These responses indicate a number of shortcomings. One is that differences between the judged frequencies of the most and least frequent events were considerably smaller than the corresponding differences in the statistical estimates. The former varied over three to four orders of magnitude, the latter over six. A further shortcoming noted by Fischhoff et al. (1978) was that relative to this primary bias (the flatness of the best-fit line or curve), there were also substantial *secondary* biases. These resulted in large differences in the estimated frequency of events compared with statistical frequencies. For instance, accidents were judged to cause as many deaths as diseases, whereas the latter actually take about fifteen times as many lives. Similarly, frequencies of death from botulism, tornadoes and pregnancy were also overestimated.

Fischhoff et al. (1978) argued that this last pattern of responses could be related to one of the most general judgemental heuristics: *availability*. People who use this heuristic judge an event as likely or common if instances of it are relatively easy to imagine or recall. Frequently occurring events generally come to mind more readily than rare events. Thus, quite often availability is an appropriate cue. However, availability is also affected by numerous factors unrelated to frequency of occurrence. For example, a recent plane crash or train disaster can have substantial distorting effects upon risk judgements.

Closer inspection of their data revealed that the frequencies of events such as accidents, tornadoes, floods and cancers were most overestimated, whereas most underestimated were events such as deaths from smallpox vaccination, diabetes, lightning, stroke and tuberculosis. Slovic, Fischhoff and Lichtenstein (1979) argued that these most over- and underestimated frequencies again illustrated the availability bias. Overestimated items tended to be dramatic and sensational whereas underestimated items were less spectacular events that claim one or a few victims at a time and are also common in non-fatal form. Not surprisingly, Combs and Slovic (1979) found that overestimated hazards also tended to be disproportionately mentioned in the news media.

Later studies revealed a similar pattern for estimated fatalities from various technologies. Again, low frequencies were overestimated and high frequencies were underestimated. Fischhoff, Slovic and Lichtenstein (1979) evaluated these findings by arguing that lay people's performance in these tasks is about as good as one would expect, given that they were neither specialists in the hazards considered nor exposed to a representative and adequate sample of information. Daamen, Verplanken and Midden (1986) also concluded that lay people are quite capable of ranking technological and other hazards. They also obtained systematic biases in estimation and stressed the importance of context effects (e.g., arrangement of the task, response modes). A similar point was made by Fischhoff and MacGregor (1983); they found substantial effects of response modes. For instance, asking for annual fatalities leads to answers different from those obtained when risk estimates are requested. In their case response mode did not affect the ordering of risks but had a substantial impact on absolute estimates of lethality.

It can be concluded that lay people can assess annual fatalities if they are asked to, and that they generally produce estimates similar to the statistical records. It seems, however, that their judgements of 'risk' are related to other characteristics of the hazard. These include aspects such as catastrophic potential, dramatic impact and newsworthyness and seem related to the availability heuristic.

Two further points are worth noting. First, people tend to view current risk levels as unacceptably high for several technological activities. There seems to be a substantial gap between perceived and desired risk levels. Slovic (1987) concluded that this difference clearly suggests that people are not satisfied with the way risks and benefits are balanced in society.

Secondly, experts' judgements of risk differ systematically from those of non-experts. As was noted by Slovic, Fischhoff and Lichtenstein (1979), experts' risk perceptions correlated quite highly with records of annual numbers of fatalities; their perceptions also reflected the complete range, from high to low risk, inherent in these statistical measures. Lay people's perceptions of risk, however, were compressed into a smaller range and were not as highly correlated with annual mortality statistics. When asked about perceived risks, it seems as though experts see the task primarily as one of judging technical statistics, whereas lay people gave a judgement influenced by a variety of other factors.

In a further series of studies by Slovic, Fischhoff and their colleagues it was attempted to relate perceived risk to other characteristics, such as familiarity, perceived control, catastrophic potential, equity and level of knowledge. In these studies subjects were asked to judge a large number of technologies and risk-bearing activities on characteristics such as those suggested by Starr (1969) and Lowrance (1976). These judgements were then analysed statistically in order to discover possible underlying structures.

Fischhoff et al. (1978) asked subjects to rate nuclear energy and 29

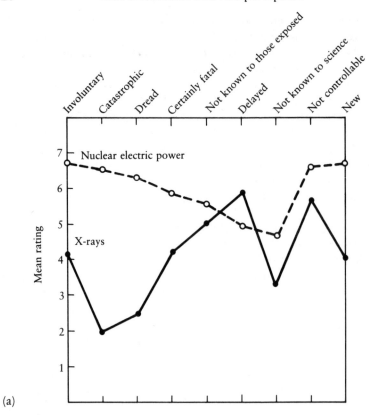

Figure 2.2 Risk profiles for nuclear electric power compared with (a) X-rays and (b) non-nuclear electric power
Source: Fischhoff et al., 1978, p. 147.

other hazardous activities on nine characteristics which were expected to influence risk perceptions. These nine characteristics were: 'voluntary – involuntary', 'chronic – catastrophic', 'common – dread', 'certainly not fatal – certainly fatal', 'known to exposed – not known to exposed', immediate – delayed', 'known to science – not known to science', 'controllable – not controllable', and 'old – new'. The thirty hazards were also rated in terms of their perceived risk.

The 'risk profiles' derived from these data showed that nuclear energy scored at or near the extreme high-risk end for most of the characteristics. Its risks were seen as involuntary, unknown to those exposed or to science, uncontrollable, unfamiliar, potentially catastrophic, severe and dreaded. The relatively unique risk profile of

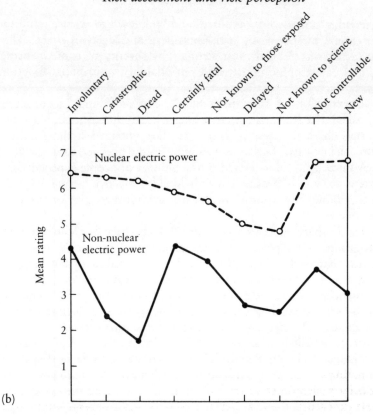

(b)

Figure 2.2 (cont.)

nuclear energy is contrasted in figure 2.2 with non-nuclear power and X-rays. Non-nuclear electric power was judged to be a much more voluntary risk, less catastrophic, less dreaded, and more familiar than nuclear energy. Nuclear energy was rated far higher on the characteristic 'dread' than any of the twenty-nine other hazards studied by Fischhoff et al. (1978). This may stem from the association of nuclear power with nuclear weapons (see Chapter 1) and from fear of radiation's invisible, potentially permanent bodily contamination, which can cause genetic damage and cancer (see, e.g., Lifton, 1976; Pahner, 1975).

Investigation of these relations by means of factor analysis indicated that the various risk characteristics could be condensed to a small set of higher order characteristics. The ratings could largely be explained by two higher order factors. The first factor is primarily determined by the characteristics 'unknown to exposed' and 'unknown to science', and to a lesser extent, by 'newness', 'involuntariness' and 'delay of effect'. The second factor was defined most strongly by severity of consequences, dread, and catastrophic potential. Controllability contributed to both factors.

In an extension of this study Slovic, Fischhoff and Lichtenstein (1980) asked 175 students to rate a longer list of 90 hazards on a total of 18 risk characteristics. In this case they obtained a three factor solution. The first two factors were similar to the ones found in the first study. A third factor was related to the number of people exposed, and the extent to which people are individually affected by the hazard. Figure 2.3 shows the location of hazards within the space of the first two factors.

The factor space presented in figure 2.3 has been replicated across various groups of lay people and experts judging large and diverse sets of hazards. Factor 1, labeled 'dread risk', is defined at its high (right-hand) end by perceived lack of control, dread, catastrophic potential, fatal consequences, and the inequitable distribution of risks and benefits. As can be seen from figure 2.3, nuclear weapons and nuclear energy score highest on the characteristics that make up this factor.

Factor 2, labelled 'unknown risk', is defined at its high end by hazards judged to be unobservable, unknown, new and delayed in their manifestation of adverse consequences. Chemical technologies score particularly high on this factor. A third factor, reflecting the number of people exposed to the risk, has also been obtained in several studies (see Slovic, Fischhoff and Lichtenstein, 1980). How do the ratings on these risk characteristics (and the two factors that summarize them) relate to the judgements of current and desired levels of risk? Perceived risk and desired magnitude of adjustment are clearly related to 'dread risk'; i.e. people overestimate and want something done about the risks they fear most. It is worth noting that experts' risk judgements tend not to correlate highly with any of the characteristics or factors. Their risk judgements are, however, closely related to technical estimates of the average annual fatalities from each activity (Slovic, Fischhoff and Lichtenstein, 1979). In other words, the experts seem to view riskiness as more or less synonymous with annual fatalities.

In a related study Vlek and Stallen (1981) asked a large sample of subjects to sort 26 activities and technologies according to the perceived similarity in their risks. These judgements were analysed with multi-dimensional scaling techniques. Their analysis revealed two factors that could explain the similarity ratings. The first concerned the size of potential accidents (e.g., high for nuclear power, low for driving a car).

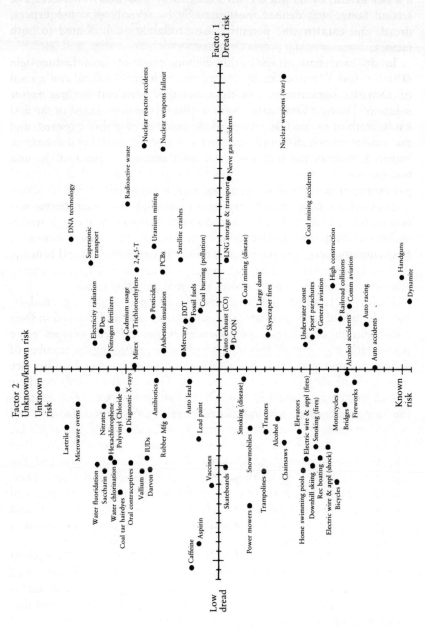

Figure 2.3 Risks of activities and technologies in a two-factor space *Source:* Slovic, 1987, p. 282. (Copyright 1987 by the AAAS.)

The second referred to the degree of organized safety (low for most individual activities, high for some technologies such as data banks). In a later study, Hohenemser, Kates and Slovic (1983) asked risk experts to rate 93 hazards on 16 dimensions. A factor analysis revealed five main factors: 'biocidal', 'delay', 'catastrophic', 'mortality' and 'global'.

Lindell and Earle (1983) related judgements of the minimim safe distance from each of eight hazardous facilities to their ratings on thirteen risk dimensions. The eight facilities were: natural gas power plant, oil power plant, coal power plant, oil refinery, liquified natural gas storage area, nuclear power plant, toxic chemical disposal facility and nuclear waste disposal facility. Their findings revealed a cluster of high-risk facilities (nuclear waste and toxic chemical waste facilities and nuclear power station). At the low-risk end, facilities included natural gas power plant, oil power plant and coal power plant.

Figure 2.4 shows the risk profiles for two high-risk facilities (nuclear power plant and toxic chemical disposal factory) and one low-risk facility (natural gas power plant) on the thirteen risk dimensions. Respondents were 396 people including environmentalists (86), urban

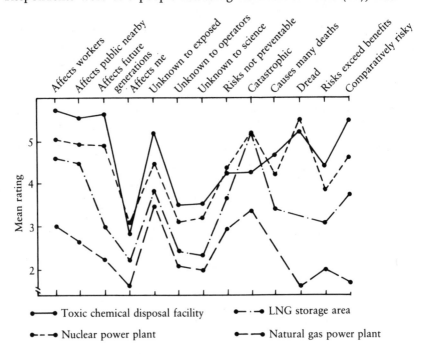

Figure 2.4 Risk profiles for selected hazardous facilities
Source: Lindell and Earle, 1983, p. 250.

residents (84), hazardous facility communities (110), science writers (38), nuclear engineers (42) and chemical engineers (36). Overall, about 78 per cent were willing to live or work within 16 km of a natural gas power plant; percentages for a nuclear power plant and a toxic chemical disposal factory were 35 and 29 respectively. Interestingly, only 23 per cent of the nuclear engineers were willing to live within 16 km of a chemical waste facility compared with 55 per cent of the chemical engineers. Percentages for a nuclear waste facility showed a reverse difference: 60 per cent for nuclear engineers, 47 per cent for chemical engineers.

The risk profiles shown in figure 2.4 are based on the total sample and show that the high-risk facilities were distinguished by high levels of perceived threat to workers, the public and future generations. Furthermore, these facilities were believed to have less well known, less preventable risks, and more catastrophic potential. They were also firmly associated with dread risk while the risk–benefit ratio indicates that the overall risks are seen as equal to or greater than the benefits derived from these two facilities. These findings can be related to those obtained by Slovic, Fischhoff and Lichtenstein (1980) and by Vlek and Stallen (1981). They would fall into the first quadrant of figure 2.3 (i.e., high on the factors 'unfamiliar' and 'dread'). Similarly they would fall into a single quadrant of the space defined by the two dimensions reported by Vlek and Stallen (1981); i.e., 'size of a potential accident' and 'degree of organized safety'.

Results of these taxonomies are useful for clarifying the structure of technological risks and could also help to understand public reactions and help predict future acceptance and rejection of specific technologies. An obvious case in point is nuclear energy. Its isolated position in the factor space reflects the public's view that nuclear energy risks are unknown, dreaded, uncontrollable, inequitable, catastrophic and likely to affect future generations. It has been argued that people's strong fears of nuclear energy and their political opposition to it are not irrational, but are logical consequences of their concerns about these considerations; i.e. equity, catastrophic potential, and the safety of future generations (cf. Slovic, Lichtenstein and Fischhoff, 1979). Furthermore, it seems likely that accidents occurring with unknown and potentially catastrophic technologies will be seen as indicative of our loss of control over the technology. Moreover, even a small accident occurring to a technology in the unknown and dread quadrant of the risk space is likely to have much greater impact than an accident causing similar immediate damage that occurs with a technology with different characteristics. Thus even 'small' accidents in the nuclear industry are likely to have considerable impact, a fact that should have direct implications for the setting of safety standards.

Let me give two examples of routine emissions and/or small accidents which have confirmed the concern of the general public (as far as they knew about them). The first is an accident at Sellafield and the possible consequences of routine emissions (a reported increase of leukaemia cases). This accident will be discussed in several chapters in the book (see, e.g., Chapter 6). Another example concerns a longitudinal study conducted in the USA. Bertell studied environmental influences on the survival of infants of low birthweight in Wisconsin between the years 1963 and 1975 (Bertell, Jacobson and Stogre, 1984). They chose six regions based on their proximity to nuclear power stations; three were close to and downwind of nuclear power stations, three were distant from or upwind of nuclear power stations. Bertell et al. studied the effects on such infants because they are very fragile and highly sensitive to environmental factors that might have negative consequences for human health. Their results showed that before the start-up of nuclear reactors the death rates declined in both sets of regions, most probably as a consequence of advances in medical technology and health care. This decline continued in the remote or upwind regions after 1970. In the downward regions, however, the death rates rose again after 1970, which corresponded with the start-up of the seven nuclear reactors between 1969 and 1974. In the period 1971–75 death rates per 1,000 infants were 175.8 for the downwind regions and 141.3 for the remote or upwind regions. As has been argued, not only major accidents but also minor accidents or reports about the effects of routine emissions are bound to have an impact on public attitudes.

Risk analysis: facts and values

The rationale for using any of the taxonomies described in the previous section in policy decisions about technologies is the assumption that risk and risk acceptability are fundamental characteristics of a technological conflict. This view also suggests that risk acceptance is crucial to resolving the conflict about nuclear energy. Although the taxonomies help by defining generic risk characteristics with differing levels of public acceptability and social conflict, it seems too simple to reduce the nuclear energy issue to a conflict about risks. This assumption was made in the early stages of risk perception research. It was hoped that psychologists could help to understand why the public did not understand risks and then formulate communication strategies and thus resolve the conflict. This, obviously, did not happen. This could be due to the psychologists' failure to explain risk perception and/or their inability to design adequate communication programmes. Another approach is to try and analyse the nuclear conflict more carefully and look at the various *levels* of conflict about technologies such as nuclear power.

Von Winterfeldt and Edwards (1984) examined patterns of conflict about risky technologies. Both these authors and Otway and von Winterfeldt (1982) argued that many technology debates go beyond the notion of risk and involve a much wider domain of political, social and ethical concerns. Similar points were made in our own work (see, e.g., Eiser and van der Pligt, 1979; van der Pligt, van der Linden and Ester, 1982). Sociological studies also provide evidence for the relevance of other factors, including the institutional, political and cultural context (see, e.g., Douglas and Wildavsky, 1982; Nelkin, 1979; Mazur, 1981).

For a long period debates about nuclear energy were technical debates about technical matters. This context emphasized facts and procedural aspects, while underemphasizing value conflicts. This distinction, however, is crucial to most technological conflicts. Institutional arrangements and processes shape both the content of a controversy and the *frame of reference* within which the controversy should be resolved. We will come to this issue later in this book. It is clear, however, that the distinction between fact and value conflicts is crucial in our understanding of the nuclear debate.

Von Winterfeldt and Edwards (1984) present six categories of conflict about risky technologies. All of these can be applied to the nuclear issue. These six categories can be summarized as follows:

(1) conflicts about data and statistics
(2) conflicts about estimates and probabilities
(3) conflicts about assumptions and definitions
(4) conflicts about risk–cost–benefit tradeoffs
(5) conflicts about the equity of the distribution of risks, costs and benefits
(6) conflicts about more general social values.

The first three levels of conflict are all relatively technical. Disagreements on these levels are usually between experts of regulatory agencies, special interest groups and other (expert) stakeholders. For instance, conflict at the first level exists when involved parties disagree about data and other statistics. Sometimes this is the result of looking at different subsets of data, sometimes of disagreement about the appropriateness of statistical techniques. For instance, the increase of leukaemia cases around Sellafield in Northwest England was the subject of a long-lasting dispute. Nuclear proponents argued that one should look at the total number of cases in the county of Cumbria and compare this with the figures for other counties, while opponents insisted on comparing the number of cases in specific villages with those in other villages. Later Harnden (1989) put forward the view that more likely causes of leukaemia should be taken into account when trying to explain the leukaemia clusters in young people that have been found near Sellafield and Dounreay (another UK nuclear establishment), and proposed that a sudden population influx can also be associated with an increased leukaemia incidence.

Another view relates the leukaemia cases to exposure to radiation of the fathers working at Sellafield. Gardner (1990) found a significant relationship between the exposure to radiation of fathers and the likelihood of the development of leukaemia in their offspring.

Conflicts at the second level involve expert opinions that cannot be substantiated by empirical data. Examples are numerous. There has been disagreement about probabilities of a serious accident at a nuclear reactor, or about probabilities of leakages from a nuclear waste repository. Disagreements about estimates concern not only the impact of nuclear accidents but also, for example, the costs of dismantling or decommissioning nuclear power plants and the cost of storing nuclear waste. Conflicts at all levels can frequently be traced to conflicts about assumptions and definitions. Assumptions about economic growth have had a major impact on nuclear power programmes (most notably in France). Similarly, a definition of risk in terms of expected annual mortality leads to the conclusion that the generation of nuclear energy is a low-risk activity. But, as von Winterfeldt and Edwards (1984), Renn (1982) and Fischhoff et al. (1978) stressed, opponents of nuclear energy tend not to agree with such a definition. They tend to focus on the largest plausible accident but would also like to include other possible negative consequences for both public health and the environment.

Conflicts at the first three levels can often be solved by discussion among experts, by data collection or by research. Unfortunately expert opinions are not only based on facts but also on intuitions. Experts sometimes have vested interests in technologies or in opposing them and this often leads to diverging views. Consultation of independent experts might help in such circumstances. Solving problems about assumptions and definitions requires explication of these assumptions and definitions which might help in reaching consensus. Consensus among experts is sometimes difficult to reach, but as we will see in the next paragraphs, conflict resolution at the remaining three levels is often even more complicated.

The fourth level of conflict described by von Winterfeldt and Edwards (1984) includes disagreements about risk–benefit tradeoffs when evaluating possible technological developments. For instance, proponents of a nuclear waste repository may stress the benefits for both the local community (e.g., jobs) and the national economy. Opponents are more likely to doubt whether the jobs will indeed go to local people and will focus on local disruption because of building activities and workers coming into the area. Even more complicated are conflicts about how far one should go in reducing risks and at what cost. The essential tradeoff is here one of money (the costs of increased safety) versus the expected reduction in risk (e.g., expected reduction in fatalities). These tradeoffs are complex and are often accompanied by considerable conflict not only in the context of technological conflicts about nuclear energy but also in the context of medical decision-making (e.g., financial investments versus expected number of lives saved).

Related to conflicts at this fourth level are inequitable distributions of risks and benefits. Inequities usually occur if some people (e.g., a small community) are or feel heavily exposed to the risk, while others get the benefits of a technology. This issue plays a crucial role in the siting of nuclear power stations and nuclear waste facilities. (Chapter 5 focuses on the siting of nuclear installations.)

Finally, the most fundamental and most complicated level of conflict concerns basic social values. For instance, the nuclear debate is also related to issues of economic growth, degree of centralization and the scale of industrial facilities. Conflicts at the last three levels are not likely to be solved by research or consensus among experts. Moreover, non-experts are much more involved in these conflicts than the conflicts at the first three (more technical) levels. In value conflicts, economic, institutional and political aspects play more important roles.

Conclusions

In this chapter I have briefly described the roles of risk assessment techniques in formulating nuclear policy. Risk assessment could help to develop alternative policies. The newness of nuclear technology and the lack of empirical data make it inevitable that 'intuition' and 'educated guesses' play an important role in risk assessment. The resulting uncertainty leaves room to disagree about the viability (i.e., the risks and benefits) of policy alternatives.

Risk perception research has developed a number of taxonomies of technologies and has helped to clarify the structure of technological risk and factors that determine public acceptance of these risks. For example Slovic, Fischhoff and Lichtenstein (1982) put forward the view that the controversies surrounding nuclear power can at least partly be explained by the extreme position that nuclear risks assume in their factorial representations. Nuclear power is seen as unknown, dreaded, uncontrollable, catastrophic, inequitable and likely to affect not only the present but also future generations.

However, closer inspection of the conflicts that exist in the nuclear debate reveals a variety of disagreements. Some are very factual, others concern social issues such as equity or more basic social values such as the desirability of economic growth. Following von Winterfeldt and Edwards (1984) I described six levels of conflict. These show that the nuclear debate is not only about risk perception or risk acceptance. Risk–benefit issues, equity and social and political values all play important roles in the nuclear issue. In the next chapters I will describe attitudinal research that attempts to widen the scope and also include some of these other sources of conflict. Chapters 3 and 4 focus on general and local attitudes towards nuclear power and nuclear power stations. In Chapter 5, I will describe the process of siting nuclear facilities and focus on the equity issue.

3
Attitudes, beliefs and values

Expert assessment of the risks of nuclear energy indicates that these are no greater than the risks associated with other generally accepted technologies – indeed, they may be smaller. The public, however, has a substantial distrust of nuclear energy. As Chapter 1 showed, opinion polls consistently report public worries about the release of radioactivity, possible catastrophic accidents, and the safe disposal of nuclear waste. Possible adverse consequences for health and the environment are seen as major risks of nuclear energy.

Although risk perception plays an important role in public acceptability of nuclear energy, the concept of risk does not embrace all the relevant aspects of public acceptance. As we saw in the previous chapter, conflicts about nuclear energy concern a variety of facts and values over and above the concept of acceptable risk. The public's perceptions of risks are built on values, attitudes and sets of attributes which need not be similar to those of risk experts and policy-makers. Research on public attitudes should result in a better understanding of the factors that determine overall acceptability of nuclear energy.

This chapter deals with these issues. First, I will introduce the attitude concept and discuss the concept of dimensional salience. Next, I will present some of the empirical research on these issues in more detail. Finally, I will turn to processes that play a role in the perseverance and polarization of attitudes towards nuclear power.

Attitudes

The concept of attitude has a long history in social psychology and, as was noted by Dawes and Smith (1985), it has known many definitions.

There is little agreement about the definition of attitude. As early as 1935 Allport observed that 'attitudes are measured more successfully than they are defined' (p. 9). Two classes of definitions have dominated attitude research since around 1960. The first stresses affective aspects of attitude and was introduced by Katz (1960). Katz wrote that 'Attitude is the predisposition of the individual to evaluate some symbol or object or aspect of his world in a favorable or unfavorable manner. Attitudes include the affective, or feeling core of liking or disliking, and the cognitive, or belief elements which describe the effect of the attitude, its characteristics, and its relation to other objects' (p. 168). The second approach extended this definition to include behavioural intentions. This 'three-component' view of attitudes, proposed by Rosenberg and Hovland (1960), has provided a general framework for the study of attitudes. In this approach, attitude is assigned the status of an intervening variable between 'stimuli' (objects, people, events) and 'responses' to these stimuli. An attitude is defined as having three components: affect (concerned with feelings, evaluations and emotions); 'cognition' (concerned with beliefs about whether something is true or false); and 'behaviour' (concerned with behavioural intentions). Attitudinal research on the nuclear issue has paid little attention to the behavioural component of attitudes and tends to focus on the affective and cognitive elements.

The models most often used to study the structure of attitudes towards nuclear energy issues are all related to expectancy-value approaches of the attitude concept. These approaches are based on the early work of Edwards (1954) on 'subjective expected utility'. The two major components of subjective expected utility theory are anticipated costs and benefits. Expectancy-value models of attitude are a close parallel to the expected cost-benefit-utility relationship in the decision-making literature. The work of Fishbein and his colleagues has been very influential in the attitude literature of the last two decades (Fishbein and Ajzen, 1972; 1975; Fishbein and Hunter, 1963).

The fundamental proposition of their expectancy-value model states simply that 'a person's attitude towards any object is a function of his or her beliefs about the object and the implicit evaluative responses associated with those beliefs' (Fishbein and Ajzen, 1975, p. 29). Algebraically, the attitude towards an object

$$A_o = \sum_{i=1}^{n} b_i e_i$$

where b_i is the belief i about the object 'o' – that is, the subjective probability that there is a specific association between 'o' and some other object, concept, value or goal 'x_i'; e_i refers to the evaluation of b_i with n being the number of beliefs a person holds about the attitude object 'o'. Alternatively stated, the attitude towards the object is a function of the subjective probability that the object is related to attribute i multiplied by the evaluation of that specific attribute, summed over the relevant attributes.

Quite often these attributes will be anticipated outcomes that could occur in relation to the attitude object in question. As can be seen from this brief overview, there is a close relationship between the notions of anticipated cost and benefits, the subjective expected utility concept and Fishbein and Ajzen's model. The sum of the subjective probabilities weighted by their evaluation can be viewed as a measure of overall utility of an attitude object. Benefits are simply the sum of the $b_i e_i$ products of positive attributes; costs are the summed products of the negative attributes.

The assumption of the model is that attitudes are the result of a 'rational' process; i.e., beliefs are formulated in terms of possible positive and negative outcomes and integrated into an overall evaluation. In other words, each individual is presumed to engage in a (simple) form of cost–benefit analysis. Although the various components are subjective, relatively crude and non-quantitative, the process for reaching an overall attitude is assumed to be rational.

This expectancy-value approach has frequently been applied to the issue of nuclear energy. In the next section the major findings of this research will be discussed.

Attitudes towards nuclear energy

Since the mid 1970s several studies have attempted to analyse the structure of people's attitudes towards nuclear energy. Two early examples are the studies by Otway and Fishbein (1976) and by Otway, Maurer and Thomas (1978). Both are applications of Fishbein's model of attitude formation, which basically assumes that the more a person believes the attitude object is associated with good rather than bad attributes or consequences, the more favourable his or her attitude will be.

Otway and Fishbein (1976) conducted a pilot study on thirty members of an American institute for energy-related research. Both pro- and anti-nuclear subjects seemed to agree about the risks of nuclear power. Their findings suggested that 'differing attitudes towards nuclear power may be primarily determined by strongly differing beliefs about its benefits' (p. 13) The particular group used in this study was not representative of the general population. It was composed largely of professional people employed at a university institute engaged in energy research. The same set of underlying beliefs would thus not necessarily have the same relevance to another group.

The most important conclusion of their study was that even with small sample sizes an analysis of the cognitive structures underlying attitudes towards nuclear energy could identify the factors differentiating between people with favourable and unfavourable attitudes. For example, the 'pro' and 'anti' attitude groups did not differ greatly in their beliefs about risk-related attributes. The major difference in the attitudes of these groups towards nuclear power was related to differing beliefs about benefits. Consistent with this finding, the benefit-related attributes were most important for the pro-nuclear group while risk-related aspects were most important for the anti-nuclear group. The main factors distinguishing between the 'risk' and the 'risk averse' groups concerned beliefs that people are exposed to nuclear risks involuntarily and in a passive way. People strongly holding these beliefs about nuclear power tended to view the risks as being unacceptable.

Otway and Fishbein (1976) relate their approach to the assumption that much of the conflict surrounding nuclear power is due to different perceptions. They further argue that the major advantage of an expectancy-value approach is that it allows relatively objective measurement and description of these perceptions through the identification of attributes used by different groups to characterise the same technology. Knowledge about the ways in which different groups of people characterise nuclear energy technology could thus provide useful information for the decision-maker (cf. Otway and Fishbein, 1976). In a later study, Otway, Maurer and Thomas (1978) report the results of a factor analysis on 39 belief statements about nuclear power based on a sample of 224 Austrians. This factor analysis yielded four factors designated as 'psychological risk', 'economic and technical benefits', 'sociopolitical risk', and 'environmental and physical risk'. Subgroups of the 50 most pro- and 50 most anti-nuclear respondents were then compared. In this way it was attempted to determine the contributions of beliefs and evaluations within each of these factors to respondents' overall attitudes. For the pro group, the economical and technical benefits factor made the most important contribution; for the anti group, the risk factors were more important. The anti group believed that environmental and physical risks would be increased by the use of nuclear energy, while the pro group believed they would not. Otway, Maurer and Thomas (1978) infer that 'Beliefs about the benefits of nuclear power appear to be relatively independent of beliefs about the risks. Further, people differentiate between types of risk' (p. 115).

The finding that separate dimensions of the issue appear differentially salient (in terms of their contributions to the prediction of overall attitude) for the different attitude groups has important practical implications for communication and mutual understanding between the protagonists in the nuclear debate. Moreover, at a theoretical level it demands a conception of attitudes which takes such differential salience into account. Another theoretical question concerns how specific beliefs and evaluations are related to global attitudes. Fishbein and Ajzen's model seems to imply that the direction of causality is from the molecular to the global, but there is also evidence that people will tend to bring their attitudes and predictions of possible future events into line with more specific, existing beliefs. For example, both the literature on cognitive dissonance (Festinger, 1957) and related literature on the selective exposure hypothesis (cf. Freedman and Sears, 1965; McGuire, 1969) would argue that the number of positive and negative beliefs held by people with different attitudes will be in accordance with their general attitudes. Moreover the literature on schemata and the confirmation bias also suggests that existing attitudes and/or implicit theories are likely to affect seriously information search, integration and memory (see, e.g., Fiske and Taylor, 1984). If this is so, then the understanding of why people hold different attitudes towards nuclear energy may require an investigation not only of their specific beliefs but also of their more general systems of values. Before discussing some of the findings in this research area I will turn to the concept of 'dimensional salience'.

Dimensional salience

When people judge attitudes and/or opinions they will discriminate more between these in terms of attributes they consider *important* or *salient* than those considered non-salient (van der Pligt and Eiser, 1984). A given dimension can be said to be particularly salient for individuals if they give particularly extreme or polarized judgements along it. Tajfel and Wilkes (1963) provided an operational definition of salience based on the criteria of priority and frequency of mention. The point here is that one can decide which dimensions are to be treated as salient for a given individual *before* looking at his or her judgements. In an attempt to provide an answer to the question of *why* a particular dimension or stimulus attribute is more salient than another, Tajfel related salience to 'value'. According to accentuation theory, the more the differences along a specific dimension are systematically related to differences in value, the more salient that dimension is. This implies that a salient dimension is a dimension that is relevant to an individual's evaluations of the stimuli he or she is required to judge. The notion that individuals discriminate between objects of judgement in terms of those dimensions they see as salient is basically a statement about *how* people attempt to interpret their environment.

As was discussed in the introduction to this chapter, several studies have shown that separate dimensions of the nuclear energy issue appear differentially salient for the different attitude groups. First I will discuss some studies relating dimensional salience of both specific beliefs and more general values to attitudes towards nuclear energy.

Salience and values

The first study was carried out at a workshop on nuclear energy held shortly after publication of a report on a proposed nuclear fuel reprocessing plant (Eiser and van der Pligt, 1979). On 14 June 1977 a public inquiry opened into an application by the state-owned company British Nuclear Fuels Ltd to expand the nuclear facilities at Windscale (now called Sellafield) in Northwest England by building a thermal oxide fuel reprocessing plant (THORP). This would reprocess spent nuclear fuel from both British and foreign nuclear reactors. Reprocessing would enable the industry to extract plutonium and make it available as fuel for reactors (particularly the new 'fast breeders'). The plutonium could then be sold back to the countries from which the spent fuel had come.

Even after reprocessing there would be highly radioactive waste products in need of careful storage (in addition to less radioactive effluents which would be discharged into the sea). The military importance of plutonium raised fears of 'nuclear terrorism' and of the international proliferation of nuclear power capability. On 6 March 1978 the report of the inquiry was published (Parker, 1978). It recommended that the development of the plant should go ahead. The House of Commons approved the report and approved a development order for the plant on 15 May.

Eiser and van der Pligt (1979) assessed individual attitudes towards this development but also presented subjects with a list of eleven possible consequences (shown in Table 3.1). They asked subjects to rate how each of these would be affected if the development of THORP went

Table 3.1 Pro- and anti-nuclear respondents' perception of consequences of expansion of nuclear industry

	Mean scores[a]		Percentage[b]	
	pro Ss (N = 25)	anti Ss (N = 19)	pro Ss (N = 24)	anti Ss (N = 18)
(a) Strength of the UK economy	1.2	0.3*	75	17*
(b) Risk of nuclear terrorism	0.3	1.3*	4	72*
(c) Ordinary UK citizen's influence over political decisions	0.2	−0.9*	8	50*
(d) Level of unemployment in the UK over the next 10 years	0.0	−0.1	50	6*
(e) Risk of a serious nuclear accident in the UK	0.2	1.2*	21	56*
(f) Restrictions on individual civil liberties in the UK	0.3	1.3*	17	83*
(g) UK's ability to meet future energy demands	1.7	0.5*	92	22*
(h) Risk of proliferation of military nuclear capability	0.2	1.4*	8	67*
(i) Total deaths among workers in UK nuclear, coal, oil and gas industries *combined* from accidents and occupational deseases	−0.5	0.2*	42	6*
(j) Total environmental damage produced by UK nuclear, coal, oil and gas industries *combined*	−0.4	0.6*	71	39*
(k) Total health hazards to members of the public from routine pollution by UK nuclear, coal, oil and gas industries *combined*	−0.5	0.7*	67	39*

[a] Possible range of scores from −2 (greatly decreased) to +2 (greatly increased).

[b] The columns do not add up to 500 because of the inclusion of 4 pros and 4 antis who chose fewer than 5 consequences.

* Scores that differed significantly (at the 0.005 level) between the two groups.

Source: Adapted from Eiser and van der Pligt (1979), p. 528.

ahead, in terms of five categories from greatly decreased (−2) to greatly increased (+2). They then had to rank the five possible consequences they thought were most important. Subjects were also asked to rank five factors which they felt 'would contribute most to an improvement in the overall "quality of life" as you understand it'. First, I will discuss the perceived consequences of THORP.

Table 3.1 presents the mean ratings by the pro and anti group of the eleven possible consequences of the THORP development. Apart from item (d), the level of unemployment, which neither group felt would be affected either way, all items produced substantial differences between the two attitude groups, in a direction predictable from a simple consistency notion (e.g., Rosenberg, 1956) as well as from Fishbein's model. The pro group, as compared with the anti group, saw THORP as more likely to produce benefits and less likely to lead to adverse consequences. We next computed a 'utility' score for each subject by multiplying the ratings (a, c, g) by +1 and the eight adverse consequences (b, d, e, f, h, i, j, k) by −1. Omitted items were scored 0. This 'utility' score correlated strongly (0.75) with the attitude measure.

Next we considered which possible consequences were chosen as being among the five most important by the two attitude groups. Table 3.1. also shows the percentages of pro and anti subjects choosing each of the consequences anywhere among the most important five. There were striking differences between the two groups, with the pro subjects most frequently choosing the strength of the UK economy and the UK's ability to meet future energy demands and the anti subjects choosing restrictions on civil liberties and the risk of nuclear terrorism. These differences in emphasis are underlined by correlational analyses. The rank correlation between the two sets of percentages was −0.50. We then recalculated subjects' 'utility' scores for only those items which they selected among the five most important. The 'utility' scores based on the selected important items correlated 0.86 with subjects' attitudes. A similar calculation on the six less important items for each subject yielded scores which correlated only moderately (0.44) with subjects' attitudes.

These findings indicate that the two groups had different reasons for holding their attitudes. Moreover, focusing on the consequences they found important resulted in a significantly higher correlation between the overall attitude measure and the more specific beliefs about possible consequences of expanding the nuclear facilities at Sellafield.

A similar study was conducted in the Netherlands (van der Pligt, van der Linden and Ester, 1982). A sample population (600 people) from four communities at varying distances from a nuclear power station

Table 3.2 Estimated likelihood and importance of potential consequences of nuclear energy as a function of attitude

	Mean score[a]			Percentage[b]		
	Pro subjects (N = 179)	Neutral subjects (N = 72)	Anti subjects (N = 349)	Pro subjects (N = 179)	Neutral subjects (N = 72)	Anti subjects (N = 349)
(a) Increasing the strength of the Dutch economy	1.3	1.7	2.7[c]	5	25	9
(b) A serious nuclear accident in The Netherlands	3.0	2.2	1.3	7	21	68
(c) Securing The Netherlands' future energy demands	1.7	1.9	2.9	4	33	8
(d) Serious health consequences due to storage of nuclear waste	2.5	1.7	1.2	5	35	86
(e) Reduction of the level of unemployment in The Netherlands	2.3	2.8	3.5	5	18	6
(f) Restrictions on individual civil liberties due to extensive security measures	3.0	2.4	2.2	2	29	20
(g) Increasing the independence of other countries	1.4	1.7	2.8	6	42	10
(h) Negative consequences for the environment	3.1	2.2	1.3	2	36	74

[a] Possible range of scores from 1 (very likely) to 4 (very unlikely).
[b] The scores represent the proportion of subjects selecting each factor among the three most important. The columns do not add up to 300 because of the inclusion of subjects who chose fewer than 3 consequences.
[c] All differences between the three attitude groups were significant at the 0.05 level.
Source: Adapted from van der Pligt, van der Linden and Ester (1982), p. 225.

answered a questionnaire on the issue of nuclear energy. Again we assessed subjects' attitudes towards nuclear energy and presented a total of eight possible consequences of a further expansion in the number of nuclear power stations (see table 3.2).

The sample was split into three groups on the basis of respondents' answers to the question concerning the building of more nuclear power stations in the Netherlands: (1) those strongly or moderately opposed to building more nuclear power stations; (2) those who were undecided/neutral; (3) those who were strongly or moderately in favour.

Table 3.2 presents the mean ratings by the pro, neutral and anti group of the eight possible consequences of nuclear energy. Results indicate that the probability estimates of the possible consequences are related to subjects' attitudes towards nuclear energy. The pro group estimated that the economic benefits of nuclear power would be considerable and also believed that the development of nuclear energy would result in more independence from other countries. The anti group saw nuclear power as more likely to produce serious accidents with adverse consequences for both the environment and public health. Finally, the anti group thought it more likely that the storage of nuclear waste would become a major problem.

We next computed an index score along the lines of Fishbein's (1963) expectancy-value model. These $\Sigma b_i e_i$ scores (probability ratings, weighted by evaluation, summed over items) were calculated by taking into account only the presumed sign and not the degree of evaluation. In other words, the ratings of the benefits were multiplied by +1 and those of the negative consequences by −1. The score thus obtained correlated 0.80 with subjects' ratings on the scale 'extremely opposed – extremely in favour' on increasing the number of nuclear power plants, and −0.68 with subjects' attitudes towards closing the existing nuclear power plants.

Table 3.2 also shows which possible consequences were chosen as being among the three most important by the three attitude groups. The results provide strong support for the findings described earlier in this chapter; i.e., pro subjects regard economic aspects of nuclear energy as most important, while antisubjects rate as most important the risk of nuclear accidents and the adverse consequences for the environment. In addition, the undecided group considered *both* economic and environmental aspects to be relatively important.

In this study we also calculated subjects' $\Sigma b_i e_i$ score for only those beliefs which they individually selected among the three most important. The correlation between these scores and subjects' attitudes towards increasing the number of nuclear power plants was highly significant

Table 3.3 The importance of general values for pro- and anti-nuclear respondents

	Percentage of respondents selecting each factor	
	Pro subjects	Anti subjects
(a) Decreased emphasis on materialistic values	36	100
(b) Higher material standard of living	36	0
(c) Reduction in scale of industrial, commercial and governmental units	22	86
(d) Greater public participation in decision-making	50	66
(e) Industry modernization	68	6
(f) Advances in science and technology	82	13
(g) Security of employment	77	40
(h) Improved social welfare	31	80
(i) Conservation of the natural environment	77	100

Source: Adapted from Eiser and van der Pligt (1979), p. 532.

(0.83). A similar calculation on the remaining (less important) consequences resulted in a significantly lower correlation (0.65).

The findings of both of these studies also suggest that the different perceptions of the possible consequences of further expansion of the nuclear industry are related to more general values. To investigate this issue we asked the forty-seven participants attending the previously mentioned one-day workshop to select the five factors which they felt 'would contribute most to an improvement in the overall "quality of life"' from a list of nine. Table 3.3 summarizes the findings.

Results show substantial differences between the two groups, with pro subjects stressing the importance of 'advances in science and technology', 'industrial modernization', 'security of employment' and 'conservation of the natural environment'. The anti-nuclear respondents put even more emphasis on the last factor and stressed the importance of 'decreased emphasis on materialistic values' and 'improved social welfare'. In a related study we presented a sample of the Dutch population with a similar list of more general values (van der Pligt, van der Linden and Ester, 1982). Results were in accordance with the above study and showed that pro-nuclear respondents stressed the importance of economic development, whereas anti-nuclear respondents put more emphasis on conservation of the natural environment and the reduction

Table 3.4 Importance of various social issues as a function of attitude

	Pro subjects (N = 179)	Neutral subjects (N = 72)	Anti subjects (N = 349)
(a) Maintaining the present material standard of living	31[a]	33	28
(b) Improvement of the strength of trade and industry	76	63	37
(c) Conservation of the natural environment	28	31	58
(d) Reduction of the level of unemployment	64	75	71
(e) A stricter criminal law	35	42	24
(f) Providing a less complex society	17	17	29
(g) Greater public participation in decision-making	3	1	12
(h) Increase of defence spending	24	11	3
(i) Reduction of energy use	16	18	26

[a] Scores represent the percentage of subjects selecting each issue among the three most important. The three columns do not add up to 300 because of the inclusion of subjects who chose fewer than three issues.

Source: Adapted from van der Pligt, van der Linden and Ester (1982), p. 226.

of energy use. Finally, the anti-nuclear group thought the issue of increased public participation in decision making to be more important (see table 3.4)

A stepwise discriminant analysis was conducted on these scores to find the factors that most distinguished the pro group from the anti group. The results revealed three factors producing a reasonable degree of separation. These factors were: 'improvement of the quality of trade and industry'; 'increase of defence spending'; 'conservation of the natural environment'.

Not surprisingly, results also showed a relation between these value differences and respondents' position on a political left–right dimension. Political preference was significantly related to these differences in values *and* to respondents' attitudes to the building of more nuclear power stations. Opinion poll surveys conducted when the second study was being carried out confirmed this relationship between political preference and attitudes towards nuclear energy (see van der Pligt, van der Linden and Ester, 1982).

The results of these two studies provide strong support for the view that individuals' attitudes towards nuclear energy are closely related to their perceptions of its potential consequences. These findings are similar to those obtained by Otway, Maurer and Thomas (1978), Sundstrom et al. (1977; 1981) and Woo and Castore (1980), and show that individuals approach the issue of nuclear energy in terms of various potential positive and negative consequences.

Results also support the view that individuals with opposing attitudes tend to see different aspects of the issue as salient. They will disagree not only over the likelihood of the various consequences but also over their importance. In other words, each group has its own reasons for holding a particular attitude: the pro groups saw the potential economic benefits as most important; the anti groups attached greater value to environmental and public health aspects. The undecided groups showed a more balanced view and rated both aspects as relatively important. A further indication of the importance of belief saliency is provided by the finding that the correlation between subjects' utility scores and their attitudes was considerably higher when considering only those consequences selected among the three most important, than when only less important consequences were considered.

Results concerning subjects' perception of the importance of more general social issues were in line with the above findings. Similar findings were obtained by Lindell and Earle (1980). These results clearly indicate that attitudinal differences towards nuclear energy are embedded in a wider context of attitudes towards more general social issues. Public thinking on nuclear energy is not simply a matter of perceptions of risk but is also related to more generic issues, such as the value of economic growth, high technology and centralization (see also Kasperson et al., 1980). It seems impossible, therefore, to detach the issue of nuclear energy from questions about the kind of society in which people want to live.

Perseverance of attitudes

The second part of this chapter focuses on processes that play a role in the perseverance of attitudes. These processes can lead to greater polarization between proponents and opponents – polarization seems to be one of the major characteristics of the nuclear debate. One mechanism is especially relevant to accounting for perseverance phenomena, i.e., *distortion* in the judgement of potentially pertinent data. That is, the weight people assign to evidence is determined, in large measure, by its consistency with initial impressions and/or existing attitudes. More specifically, the possibility that evidence seemingly consistent with existing impressions may nevertheless be irrelevant or tainted is often neglected. Similarly people too readily tend to conceive and accept challenges to contradictory evidence. As a result, data considered after the formation of a clear impression will typically seem to offer a large measure of support for that impression. Indeed, even a random sample of seemingly relevant data 'processes' in this manner may serve to enhance rather than challenge an erroneous judgement or decision. The capacity of existing impressions and expectations to bias interpretations of data is, of course, a well-replicated phenomenon in social psychology. The effects of distortion may be further enhanced by what Ross (1977) calls the autonomy achieved by distorted evidence. Once formed, an initial impression or decision may not only be enhanced by the distortion of evidence, it may even be sustained by such distortion. The autonomy enjoyed by distorted inferences may further contribute to the perseverance of non-optimal theories and decision-making strategies. Some of these biases tend to be enhanced in specific settings. The concept 'groupthink' introduced by Janis (1972; 1982) is of particular relevance here. Groupthink is a strong psychological drive for consensus within insular, cohesive decision-making groups with the result that disagreement is suppressed. This can affect the quality of decision-making when social pressures dominate the decision process at the expense of a more balanced consideration of alternatives. Janis investigated groupthink in the context of elite decision-making groups.

His analysis suggested that otherwise competent members of decision-making groups tended to overlook or simply dismiss clear indications of external threats or of flawed reasoning. Groupthink is further accompanied by symptoms such as the illusion of invulnerability, the belief in the inherent morality of the group, the illusion of unanimity, self-censorship when dissenting thoughts arise and the self-appointment of so-called 'mindguards' who try to prevent exposure to dissent. Consequences of groupthink include incomplete survey of values and/ or possible choice alternatives, poor information search, failure to reappraise rejected alternatives, and failure to work out contingency plans. Rationalizations and excessive confidence in the group's decisions are likely to make a change to more efficient decision-making methods less probable. Janis has documented both antecedents and consequences of groupthink in a number of famous policy disasters, and suggests ways to mitigate the consequences by introducing explicit counternorms (e.g., procedures that increase the likelihood of alternative views being adequately considered). It needs to be noted, however, that the list of decision defects overlaps substantially with possible shortcomings at the individual level. In other words, the phenomena are not unique to cohesive, insular, elite decision-making groups.

The preceding paragraphs suggest that many things can go wrong in attitudinal and decisional processes both at the individual and the group level. All of this has been studied extensively in other domains, but only sporadically in the domain of nuclear energy. In an earlier study (Eiser and van der Pligt, 1979) we looked at one specific aspect, i.e., the truth value assigned to various (false) statements about nuclear energy. The next section presents this issue in more detail.

Information processing: the effects of own attitude

In the study described earlier in this chapter we presented subjects with a series of factual statements (Eiser and van der Pligt, 1979). Table 3.5 shows the acceptance or rejection of arguments on either side of the issue as a function of own attitude. Results indicate relatively moderate average levels of rejection or acceptance. In all, the differences were in the expected direction and significantly so on five of the arguments. The obtained differences did not include extreme items, such as the statement about the lack fatalities caused by contact with plutonium and the statement about the explosive power of a nuclear reactor. It needs to be added, however, that the respondents were quite knowledgeable about the issue.

Survey findings show more marked differences between attitude groups when confronted with informational statements. The EC survey (Commission, 1982) indicated that 38 per cent of the total sample believed that nuclear power plants could explode like the bombs used in Japan in World War II. Fifteen per cent indicated that they did not know; only 9 per cent reported it technically impossible. Nealey, Melber and Rankin (1983) also found that a majority of the public believes that a nuclear power plant can have a massive nuclear explosion and that a small majority of the public believes that a nuclear power plant can have an atomic fission explosion (an explosion that is technically impossible). The amount of agreement with such statements is clearly influenced by one's attitude towards nuclear energy.

The same applies to the credibility and believability of specific sources. On average both pro-nuclear and anti-nuclear subjects rate university scientists as believable; differences, however, become apparent when one considers the believability of industrial scientists and the nuclear industry. Both are rated much lower by anti-nuclear subjects (Decima,

Table 3.5 Mean scores for pro- and anti-nuclear respondents on informational beliefs

	Pro subjects (N = 25)	Anti subjects (N = 20)
(a) Plutonium is the most toxic substance known to man	−0.7[a]	0.3*
(b) Apart from Nagasaki in 1945, there is no record of anyone in the world having died as a result of contact with plutonium	−0.2	−0.7
(c) Workers who actually handle plutonium are at least 7 times and possibly 20 times more likely to die of leukaemia than other workers	−0.4	0.5*
(d) There is no practical alternative to nuclear energy as a candidate for the fuel technology to succeed oil and gas	1.0	−1.4*
(e) The reprocessing of used nuclear fuel is unnecessary; it can be stored safely in the form in which it comes out of the reactors	−1.0	0.2*
(f) No reactor can explode with the force of an atom bomb	0.6	0.1
(g) Almost to a man the pro-nuclear lobby has a personal interest of some kind in the development of nuclear energy	−0.8	0.3*

[a] Possible range of scores from −2 (definitely untrue) to +2 (definitely true).
* Scores that differed significantly (at the 0.05 level) between the two groups.
Source: Adapted from Eiser and van der Pligt (1979), p. 531.

1987). Eiser, van der Pligt and Spears (1988) obtained similar findings suggesting a clear impact of own attitude on the perceived credibility of information sources and the trustworthiness of information provided by these sources. In other words, both the content of the information and the source of information receive different treatment depending on one's own attitude. The next effect of these mechanisms is that information is processed in such a way as to bolster one's existing attitudes.

Attitudes, consensus and person perception

Another way to bolster one's attitudes is to think that they are shared by many others. People seem to have a tendency to perceive a 'false consensus' with respect to the commonness of their own responses; i.e., people overestimate the commonness of their own behaviour/ attitude, relative to other behaviours or attitudes (cf. Ross, Greene and House, 1977). Such biased estimates play an important role in people's interpretation of social reality. Not only can these estimates bolster people's opinions and preferences but they also seem to be related to processes of stereotyping and attribution. Ross, Greene and House (1977) argued that individuals judge those responses that differ from their own to be more revealing of another person's personality traits than those responses which are similar to their own. In other words, people tend to ascribe relatively simplified personality traits to people holding different attitudes towards an issue. This is one way of discounting one's opponent, and can be regarded as one of the bases of stereotyping.

In one of our studies conducted in the Netherlands (van der Pligt, van der Linden and Ester, 1982; van der Pligt, Ester and van der Linden, 1983) we asked subjects with different attitudes towards nuclear energy to estimate:

(1) the percentage of the Dutch population and of the Members of Parliament in favour of increasing the number of nuclear power plants
(2) the percentage of the Dutch population that would not object to living near a nuclear power plant
(3) the percentage of residents from their local community that would not object to living near a nuclear plant.

Subjects were then asked to select from a list of twelve, the trait-descriptive terms they thought best described a typical pro-nuclear person and a typical anti-nuclear person. This list contained, in a random order, six adjectives relatively positive in evaluation (*responsible, realistic, environment-conscious, level-headed, humanitarian* and *rational*) and six relatively negative adjectives (*ill-informed, shortsighted, weak-willed, complacent, selfish* and *alarmist*). The selection was based on words used in various publications for and against nuclear energy and on material used in Eiser and van der Pligt (1979).

Results confirmed our predictions. The more favourable a person's attitude was towards nuclear energy the higher that person estimated of the percentage of the Dutch population in favour of expanding the nuclear programme. Extremely pro-nuclear respondents thought that 58 per cent of the population would be in favour, while extremely anti-nuclear subjects estimated only 29 per cent. Similarly, the first group estimated that 56 per cent of Members of Parliament were in favour, while the anti-nuclear group estimated this percentage to be 43. In other words, both groups saw their own choices and judgements as relatively common. The fact that there was no clear-cut majority for or against nuclear energy in Parliament did not prevent biased estimates, even though this had received widespread attention in the media. It needs to be noted, however, that the effect was less pronounced for estimates of MPs than for estimates of the general population. The same effect was observed with various other measures (e.g., estimates of the number of people in one's community who would object to living near a nuclear power plant).

The false consensus effect has been reported in over 45 published papers (see Marks and Miller, 1987), all confirming this tendency to see one's own choice as relatively common. Ross, Greene and House (1977) propose that this bias also implies that people see their own choice as appropriate to existing circumstances while viewing alternative responses as uncommon, deviant or inappropriate. Next, we will investigate whether this was indeed the case; i.e., did people attribute more traits to others with differing views than to those with opinions similar to one's own viewpoint?

Respondents' selections of trait-descriptive terms to describe the typical 'pro'- and 'anti'-nuclear person revealed, as predicted, a strong tendency to describe their 'own side' positively and the opposition negatively. A composite score was calculated by counting the number of positive adjectives minus the number of negative adjectives attributed to the two target persons. Results of this evaluative score are shown in figure 3.1, with subjects' attitudes towards building more nuclear power stations indicated on the five-point scale ranging from 'strongly in favour' to 'strongly opposed' on the horizontal axis. These results show clear differences in evaluation as a function of own attitude.

It is interesting to take a closer look at the ratings of the more extreme attitude groups and investigate the role of affective intensity in perceptions of consensus and person. Results show a clear relationship between extremity of attitude and consensus estimates, with higher consensus estimates for those with extreme attitudes as compared with respondents with moderate attitudes. Consensus estimates of the respondents with more extreme attitudes were not related to their willingness to infer personality characteristics. Although we did not find an effect of attitude extremity on the *number* of trait inferences, results show a clear *valence* effect in trait attributions. In other words, the groups' preferences for trait attributions seem primarily determined by evaluative factors. This effect is especially strong for those with

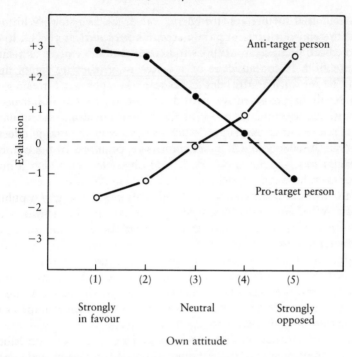

Figure 3.1 Evaluation of pro-nuclear and anti-nuclear target persons
Source: van der Pligt, van der Linden and Ester, 1982, p. 228.

relatively extreme and involved attitudes. The data also reveal that each attitude group has very different 'images' of the other.

Both mechanisms discussed in this section (overestimating the prevalence of one's own opinion and the use of what amount to personal arguments against one's opponents) could help to bolster existing attitudes and increase attitude polarization.

Conclusions

The fact that public attitudes towards nuclear energy issues are relatively stable and embedded in a wider context of values suggests that large-scale attitude conversion may be more difficult than is often assumed. People may, however, change their attitudes as a function of serious accidents that attract widespread attention, such as those at Three Mile Island (1979) and Chernobyl (1986), especially if they have not committed themselves strongly to one of the two sides. With regard to safety-related aspects of public acceptance of nuclear power, it seems much easier for nuclear attitudes to become suddenly more anti-nuclear because of a major accident or a series of smaller accidents than it would be for nuclear attitudes to become more pro-nuclear as a longer-term result of an extensive period of safe operations. Changes in a pro-nuclear direction are more likely to be gradual and to result from events related to energy supply, such as developments that would make non-nuclear energy much more expensive or more damaging to the environment.

Conflicts between opponents and proponents of nuclear energy often take the form of accusations about the other side missing important key aspects. What constitute 'key aspects' is, of course, the central question in controversies of this kind. It seems that the pro- and anti-nuclear groups tend to see different aspects of the issue as *salient*, and tend to disagree not only over the truth of certain facts but also over their importance. The correlation between attitudes and ratings of specific possible consequences was about twice as high when considering only those consequences selected by each subject as most important, as when considering only the less important consequences. Moreover, pro- and anti-nuclear attitude groups differ considerably in the consequences they see as most important. Findings discussed in this chapter indicate that economic and technical benefits contribute more to the attitudes of pro-nuclear respondents, while the health and socio-political risk dimensions contribute more to the attitudes of anti-nuclear respondents.

A similar picture emerges from perceptions of the factors contributing to an improvement in 'the quality of life'. Anti-nuclear respondents appear to be far more committed to the philosophy of 'small is beautiful' and less convinced of the benefits of technological advances and improvements in material well-being. Anti-nuclear respondents are far more preoccupied than pro-nuclear respondents by questions about the kind of society in which they want to live and the amount of power it is desirable to put in the hands of 'experts'. Pro-nuclear respondents tend to see the debate primarily as concerning the adequacy of safety precautions.

The findings presented in this chapter point to the importance of salience in the formation and maintenance of attitudes. Finally, selectivity and biases in information processing play a major role in the perseverance and, unfortunately, the polarization of attitudes. The next chapter discusses salience and selectivity in the context of attitudes towards *local* developments of nuclear facilities.

4
Community attitudes towards nuclear power stations

Chapter 3 focused on general attitudes towards nuclear power; this chapter concentrates upon local acceptance of nuclear power plants. Most research in this area employs expectancy-value models to examine the structure of attitudes towards a specific nuclear facility.

The chapter will first describe local attitudes as revealed by a study conducted in Southwest England. This study took place in the early stages of the selection of possible sites for a nuclear power station. I will then review research on local attitudes in the later siting stages and during construction. Finally, the effects of familiarity (i.e., living near a nuclear power station) upon local acceptance will be investigated.

Local versus general attitudes

A study published in 1986 (van der Pligt, Eiser and Spears, 1986a) compared attitudes towards nuclear energy in general with attitudes towards local development of a nuclear power station. The findings suggested that there is more intense opposition to possible *local* developments compared with opposition to the future development of nuclear energy in general. In the 1986 study we investigated the distribution of attitude scores towards a total of seven possible industrial developments, including nuclear power stations. Respondents (a sample of 300 residents from localities shortlisted by the Central Electricity Generating Board as possible sites for the next nuclear power station in Southwest England) were generally opposed to the construction of a nuclear power station in their neighbourhood. As shown in table 4.1, approximately 75 per cent of our respondents (categories 1, 2 and 3) indicated opposition to the proposed development in their own locality.

Table 4.1 Attitudes towards local industrial developments

Attitude towards	Very strongly opposed 1	2	3	Neutral 4	5	6	Very strongly in favour 7
More nuclear power stations in the UK	29	7	17	27	16	1	4
New nuclear power station in the locality	58	6	12	15	9	0	2
New nuclear power station in the southwest	34	7	18	26	13	2	1
Coal-fired power station in locality	38	8	18	24	11	1	2
Any industrial development taking up the same area in locality	38	7	18	22	12	3	2
Windmills for generating electricity taking up the same area in locality	22	4	12	23	28	2	10
Chemicals factory taking up the same area in locality	59	9	23	7	2	0	0

Scores give the percentage in each attitude category. Due to rounding, the percentages do not always add up to 100.

Source: Adapted from van der Pligt, Eiser and Spears (1986a), p. 6.

This percentage decreased to about 58 per cent when the proposed plant would be built somewhere else in the Southwest, while just over 50 per cent were opposed to the building of more nuclear power stations in the UK. All in all, these findings provide a clear illustration of the NIMBY effect reported in the first chapter.

Results show that the most unfavourable reactions were to the building of a nuclear power station and to the building of a chemicals factory in the locality. The most favourable reaction was towards the building of windmills. Comparisons with a more general nuclear attitude based on a series of statements indicated that attitudes towards a new nuclear power station in the locality are different from, and more extreme than, attitudes towards nuclear energy in general.

As indicated in Chapter 1, local opinion is often more pronounced than the opinions of the general public. This is likely to be related to the perceived risks and benefits in the affected communities. One of the first studies investigating local acceptance of a nuclear power station covered a period of five years, and followed local acceptance throughout the construction of the power station (Hughey, Sundstrom and Lounsbury, 1985). This study will be discussed in the next section.

Local acceptance during construction

The research by Hughey et al. (1983; 1985) took place in Trousdale County in central Tennessee. Residents in this small county were employed primarily as farmers or as workers in small factories in the only incorporated city of the county, Hartsville. The community was selected to host one of the world's largest nuclear power stations. The Hartsville nuclear power plant was planned to occupy 770 heckres and contain four reactors with an estimated generating capacity of 1,250 megawatts. The total costs were initially estimated at $2.5 billion. Estimated costs expanded dramatically, and plans for two reactors were delayed soon after construction began (cf. Hughey et al., 1985). Construction was also delayed and started in 1976. The first reactor was supposed to be constructed in five years and the entire project should have been completed by 1983. In fact, construction proceeded until late 1982 when Tennesee Valley Authority announced plans to delay indefinitely further construction. At this time the first of the four reactors was still unfinished. In the early stages construction required around 1,100 workers; peak construction involved well over 5,000 workers (approximately equal to the total population of Trousdale County).

At the time of the study by Hughey et al. (1985) these delays had not yet been announced. In 1975, before construction began, a majority of the residents (69 per cent) indicated that they would support construction of the plant. Five years later, this figure was reduced to 45 per cent. Furthermore, in 1975 over 60 per cent of the participants would have allowed construction of the nuclear power plant within 32 km of their homes. Five years later this was reduced to less than 35 per cent. Moreover, a majority (54 per cent) wanted the nuclear power station to be at a distance of 160 km or more. Hughey et al. (1985) interpreted the latter result as an indication of strongly negative local attitudes.

The results of their study thus indicated a substantial decline in local acceptance of the nuclear power plant during construction. The design of the study made it difficult to establish a direct causal link between attitudinal shifts and changes in perceived risks and benefits. Both the economic decline in the late 1970s and the accident at Three Mile Island could have played a role in the considerable reduction of local acceptance.

The study clearly showed that acceptance represented a tradeoff between benefits, costs and risk. Favourable attitudes were related to perceptions that benefits (such as more jobs and better schools) were likely consequences of the construction and operation of the nuclear power station. Adverse effects (such as health hazards and drugs in the schools) were seen as unlikely. Opponents, on the other hand, viewed health hazards and community disruption as more likely. They were also more pessimistic about potential benefits. After five years the balance in costs, risks and benefits had shifted towards seeing the plant primarily as a source of concern. Hughey et al. (1985) suggest that this could be due to changing expectations regarding the potential benefits.

In analysing the assessment of costs and benefits Hughey et al. (1985) factor-analysed the likelihood ratings of twenty-seven possible positive and negative outcomes. Results indicated five factors. Table 4.2 provides more information about the five factors obtained in their analysis.

Table 4.2 Factors underlying attitudes towards a local nuclear power plant

Factor I: Economic Growth	Factor III: General Economic Benefits
Increased local business	More entertainment
More jobs	Cheap electricity
More stores and shopping areas	Industrial development
Increased land values	Town becomes a tourist attraction
Better pay	Better schools
Housing shortage	Factor IV: Community Disruption 1
Meeting new people	Drugs in schools
More recreation areas	More taverns and bars
Recognition of town	Increased noise
More billboards	Traffic congestion
Fector II: Hazards to Safety and Environment	Factor V: Community Disruption 2
Radiation hazards	Increased taxes
Sabotage	Crowded schools
Air pollution	Increased crime
Pollution of lake	
More foggy days	

Source: Adapted from Hughey et al. (1983), p. 665.

As can be seen in table 4.2, Factor I – Economic Growth – included such possible consequences as increased local business, more jobs, and more stores and shopping areas. Factor II – Hazards to Safety and Environment – included radiation hazards, air pollution, the pollution of a nearby lake, and sabotage. Factor III – General Economic Benefits – was defined by such elements as industrial development and more entertainment. Factor IV – Community Disruption 1 – included such issues as drugs in the schools and traffic congestion. Factor V – Community Disruption 2 – included such issues as increased crime and crowded schools.

An interesting finding obtained by Hughey and his associates was that some potential outcomes were seen as more important than others. In their sample, health hazards and environmental costs were apparently more important than economic benefits and local disruptions. Sundstrom et al. (1981) obtained similar findings. Both studies supported the findings discussed in Chapter 3, and point to the role of relative importance or salience of specific beliefs. Table 4.3 summarizes the findings concerning the relative contribution of each of the five factors.

As can be seen, the second factor (Hazards to Safety and Environment) is the best predictor of acceptance, followed by 'community disruption' and 'economic growth'. Only the first three factors contributed significantly to the overall prediction of acceptance.

Local attitudes during siting

Woo and Castore (1980) examined the expectancy-value model in the context of a nuclear power plant siting controversy. In this study the perceived benefits and costs were related to attitudes towards a

Table 4.3 Multiple regression analysis of local acceptance of a nuclear power plant

Predictor	Multiple R	Multiple R^2	Change in R^2
Hazards	0.52	0.27	0.27
Community disruption (1)	0.57	0.33	0.06
Economic growth	0.61	0.37	0.05
Community disruption (2)	0.62	0.38	0.01
General economic benefits	0.62	0.38	0

Source: Adapted from Hughey et al. (1983), p. 665.

Table 4.4 Perceived consequences of a proposed nuclear power plant as a function of attitude

	Attitude		
Potential outcomes	Pro (N = 82) (%[a])	Neutral (N = 43) (%)	Anti (N = 45) (%)
Economic factors			
More electricity available	85	67	78
Slower increase in costs of electricity	44	49	33
More local jobs	61	65	69
More local business	43	53	36
Environmental factors and health risks			
Pollution (thermal, visual etc.)	72	74	96
Radiation hazard	83	56	91
Disturbances during construction	18	19	38
Too much industrial development	19	14	40

[a] Scores represent the percentage of each group of respondents who either voiced or indicated an awareness of each issue.

Source: Adapted from Woo and Castore (1980), p. 228.

proposed nuclear power plant. All participants lived within 48 km of the proposed construction site in northern Indiana. In their analyses they focused on differences between attitudinal groups (pro, neutral towards and anti the proposed power plant). The results obtained by Woo and Castore indicated that understanding the positions of the three attitude groups required two types of information. First, the number of beliefs about the possible consequences held by the attitude groups *and*, secondly, the subjective weights or importance attached to these beliefs. Table 4.4 summarizes the frequency with which the various potential positive and negative outcomes were associated with the proposed nuclear facility by the three attitude groups.

The overall, perceived utility (combined positive and negative outcomes) of the proposed nuclear power plant was significantly related to the overall evaluation of it. The pro group saw more benefits than costs; the reverse was true for the anti group. Respondents were also asked to indicate the importance they attached to the various possible outcomes. Not surprisingly, the pro groups attached much more value to the potential economic benefits, while the anti group was particularly concerned with potential health and safety issues. Furthermore, the three groups were found to differ from each other more in their perceptions of the costs than of the benefits. This could well be related to the then prevailing certainty about the economic benefits. Experts tended to disagree much more about the environmental and health risks. The next section will take a closer look at these differences in importance attached to the various potential outcomes of a local nuclear power station.

Salience and local attitudes

A series of studies attempted to investigate the relationships between people's attitudes towards the building of a nuclear power station in their locality, their specific beliefs about the immediate and long-term local consequences and their perception of the importance of these consequences (Eiser, van der Pligt and Spears, 1988; van der Pligt, Eiser and Spears, 1986a; 1986b). The research was conducted in localities that were shortlisted by the Central Electricity Generating Board (CEGB) as possible locations for a new nuclear power station in Southwest England. The sample of respondents was drawn from the electoral register for three communities which were close to the three shortlisted possible sites.

The questionnaire was closed-ended and was preceded by a short introduction describing the CEGB's announcement concerning the possible locations for the next nuclear power station in Southwest England. The questionnaire included measures of subjects' attitudes towards building more nuclear power stations in the UK, in Southwest England, and in their locality. Respondents were also asked to indicate their attitude towards various other industrial developments in their locality and, finally, to indicate their attitude towards nuclear energy in general, their involvement with the issue, and which aspects should receive most attention in a public inquiry on the possible building of a new nuclear power station.

I will first summarize the findings concerning the perception of the various potential costs and benefits of a nuclear power station in one's locality. Citizens in the shortlisted localities were generally opposed to the construction of a nuclear power station in their neighbourhood. It needs to be added, however, that most respondents were also opposed to other industrial activities, such as the building of a chemicals plant (see also earlier sections of this chapter).

In order to investigate people's assessments of the various potential costs and benefits of the proposed construction of a nuclear power station we presented subjects with two sets of fifteen potential consequences. The first set contained fifteen immediate (local) effects of the building and operation of a nuclear power station in the locality, while the second set focused on long-term local consequences. Respondents were split into three attitude groups on the basis of their answers to the question whether they were opposed to, or in favour of the building of a nuclear power station close to their community. A discriminant analysis revealed that the three attitude groups (pro, neutral and anti) differed significantly in their assessment of a variety of immediate consequences.

The aspects that were most differentially perceived concerned the total area of land fenced off, the conversion of land from agricultural use and the prospect of workers coming into the area. Opponents generally though these developments would have an adverse impact on the quality of live in the locality, while proponents expected the impact of these factors to be relatively minimal. We also asked people to select the five aspects (out of a total of fifteen immediate consequences) they thought to be the most important. Results revealed three aspects that were rated very differently as a function of own attitude. Of the pro subjects 67 per cent regarded road building an important aspect, while only 20 per cent of the anti respondents selected this item as among the five most important. A similar difference was obtained concerning the prospect of workers coming into the area (53 per cent of the pro respondents and 18 per cent of the antis). The anti respondents, on the other hand, attached greater importance than the pro respondents to the possible conversion of land from agricultural use (58 per cent versus 27 per cent).

The mean ratings by the three attitude groups of the fifteen (mainly long-term) consequences of the building and operation of a nuclear power station in their neighbourhood also showed substantial differences. Again, we conducted a discriminant analysis to find out which aspects most distinguished the three attitude groups. The results revealed three aspects which had considerable predictive power in separating the three attitude groups. These included the perceived effect of the development on one's 'peace of mind' and the effects on the environment and wildlife. The first aspect corresponds to what Otway, Maurer and Thomas (1978) called 'psychological risk', while the other two aspects are related to what these authors termed 'environmental and physical risk' (see also Chapter 3).

Respondents were again asked to select the five consequences they regarded as most important. The results showed very marked differences between the three attitude groups. The most striking difference

Table 4.5 Perceived consequences of the building and operation of a nuclear power station

	Impact[a]			Importance[b]		
	Pro respondents (N = 30)	Neutral respondents (N = 40)	Anti respondents (N = 209)	Pro respondents (N = 30)	Neutral respondents (N = 40)	Anti respondents (N = 217)
Economic factors						
Employment opportunities	8.3	7.6	6.0[c]	73	57	15[c]
Business investment	6.7	6.0	4.2	27	28	11
Environmental factors						
Wild life	4.7	2.6	1.5	40	57	67
Marine environment	5.1	3.6	2.3	13	17	38
Farming industry	4.6	3.0	1.9	17	45	56
Landscape	4.3	2.6	1.3	23	50	66
Public health and psychological risks						
Health of local inhabitants	5.0	4.4	2.6	20	29	48
Personal peace of mind	5.4	4.3	1.7	27	17	47
Social factors						
Social life in the neighbourhood	6.9	5.7	3.0	30	7	11
Standard of local transport and local services	6.8	6.3	4.9	40	27	5
Standard of shopping facilities	6.6	5.8	4.9	20	14	4

[a] Possible range of scores from 1 (consequence will affect life in the neighbourhood very much for the worse) to 9 (very much for the better).
[b] The scores represent the percentage of subjects selecting each factor among the five most important.
[c] The differences between the 3 attitude groups were significant in all cases ($p < 0.05$).

Source: adapted from: van der Pligt, Eiser and Spears (1986a), p. 9.

concerned the possible effects on employment opportunities: 73 per cent of the pro respondents selected this item among the most important, while only 15 per cent of the anti respondents considered this aspect as important. Overall, the pro respondents stressed the importance of economic benefits, while the anti respondents stressed the risk factors (both environmental and psychological risks). Table 4.5 presents a summary of these differences in the perception of the various long-term consequences and of their importance. These differences illustrate the importance of including both beliefs and salience in the conception of attitudes. Even though the attitude groups, for example, showed relatively minor differences in their evaluation of the effects of potential employment opportunities in the locality, a majority of the pro respondents found this aspect important but only a small minority of the anti respondents.

Results of this research also indicate that the major differences between the attitude groups concern the less tangible, more long-term nature of the potential negative outcomes. Moreover, the findings suggest that the perception of the *psychological risks* are the prime determinant of attitude as indicated by the very high correlation (0.80) between this factor and attitude towards the building of a new nuclear power station. Other studies (e.g., Woo and Castore, 1980) did not find such a strong relationship between psychological risks and attitude. One reason for this could be that our research concentrated on people living very close to the proposed nuclear power station (all within 8 km). Most other studies used much wider areas around proposed nuclear power stations. As will be argued in Chapter 6, this proximity to environmental hazards is likely to accentuate the role of psychological risk.

In summary, opponents and proponents of nuclear energy have very different views on the possible consequences of nuclear energy. This applies to both the general issue of nuclear energy as discussed in Chapter 3 and to the building of a nuclear power station in one's locality. The most significant difference, however, concerns the perception of psychological risks (anxiety, stress). This factor becomes more important when people are (or will be) more directly exposed to the risks, for instance when their locality is shortlisted as a possible site for a nuclear power station (see also Chapter 6).

Salience and familiarity

In another study we investigated attitudinal structure and salience as a function of the experience of actually living near a nuclear power plant (van der Pligt, Eiser and Spears, 1986b). Several surveys have compared the acceptance of a nuclear power plant among people who live near one with that of people who do not. Other studies have followed local opinion in an area where a nuclear power station was being constructed and then became operational. From these has emerged limited support for the idea that familiarity does not necessarily lead to increased willingness to accept the building of a new nuclear power station (e.g., Thomas and Baillie, 1982; Warren, 1981). As we saw before, Hughey et al. (1983; 1985) studied changes in attitudes and expectations about a nuclear power plant among residents of a small rural community over a period of five years, starting with the time of initial siting through the peak construction phase. Their study showed large negative changes in attitudes towards the plant; these were accompanied by decreased expectations of positive outcomes.

Although overall attitudes are not necessarily more favourable as a function of familiarity, some evidence suggests that people actually living near a nuclear power plant tend to underestimate relatively the risks associated with nuclear energy (Ester et al., 1983). This is most likely to be a function of experience (assuming that no accidents have taken place), but it also accords with a simple notion of dissonance reduction (Festinger, 1957). Our own findings (van der Pligt, Eiser and Spears, 1987b) suggest that this could be the case, because the relative underestimation of the risks of nuclear power stations is most pronounced when it concerns the power station in one's own locality compared with nuclear power plants elsewhere. In one of our studies (van der Pligt, Eiser and Spears, 1986b) we investigated attitudes and beliefs concerning the construction of a nuclear power station in one's locality. More specifically, we studied the effects of *familiarity* upon attitudes and upon the various potential costs and benefits of the building and operation of a nuclear power plant. In this research we also attempted to test the notion that a consideration of *both* expected costs and benefits *and* their subjective importance or salience can provide a more complete picture than could be obtained from consideration of either factor alone.

Let me first provide some background information about the context of this research. In February 1981 the British Central Electricity Generating Board announced the names of five localities to be considered as possible sites for a new nuclear power station in Southwest England. In February 1982 the CEGB ruled out two of those five sites on geological grounds. The research to be discussed here was conducted between June and October 1982 and took place in four communities. Three were the remaining sites for a possible new nuclear power station;

the other was the area around two existing nuclear power stations, Hinkley Point in Somerset. Later that year (25 August), the CEGB, surprisingly, announced that the next station would be a third station adjacent to the two existing reactors at Hinkley Point. The decision reflected the growing belief that the nuclear industry could only be developed at sites where the local public had already accepted the nuclear industry. Developments in the UK seem to aim at the creation of 'nuclear parks' with groups of power stations concentrated in a limited number of sites. Some of these sites are also expected to host future radioactive waste facilities. For instance, the Sellafield and Dounreay sites were also investigated to establish their geological suitability to house a deep repository for low- and intermediate-level radioactive waste.

In the study to be considered here, nearly 650 respondents participated. Results indicated that respondents familiar with nuclear power stations had a slightly more favourable attitude towards nuclear energy than did respondents in the other communities. Non-familiar respondents were more involved with the issue of a possible new nuclear power station than were respondents at Hinkley Point. In order to investigate people's perceptions of the various potential costs and benefits, we compared the two groups with respect to their ratings of two sets of fifteen potential consequences. Table 4.6 shows the mean ratings by familiar and non-familiar respondents of the fifteen (mainly immediate) effects of the building and operation of a nuclear power station near their community. On the basis of their location respondents were split into two groups: one group who lived close to the existing nuclear power stations at Hinkley Point in Somerset (within 8 km) and one group who lived in the communities selected by the CEGB as possible locations for a new nuclear power station in Southwest England. Results in table 4.6 show clear differences for all items. The overall picture is that non-familiar respondents are more pessimistic about the immediate impact of the building and operation of a nuclear power station.

Because the two groups also differed in their attitude towards nuclear energy, we conducted a number of analyses of covariance with familiarity as an independent variable and the attitude-index score as a covariate. Results of these analyses revealed that attitude was significantly related to the differing perceptions of the two groups. On average, all items showed highly significant effects due to attitude towards nuclear energy. All effects due to familiarity remained significant, although they were less pronounced than the effects reported in table 4.6. Table 4.6 also shows which aspects were chosen as being

Table 4.6 Perception of direct consequences of the construction and operation of a nuclear power station as a function of familiarity

	Mean score[a]		Importance[b]	
	High familiarity (N = 218)	Low familiarity (N = 430)	High familiarity (N = 218)	Low familiarity (N = 430)
Excavation for pipelines	4.3	2.6*	6	19*
Construction traffic	3.1	2.0*	42	33*
Road building	5.1	3.4*	18	24
Conversion of land from agricultural use	3.4	2.1*	37	50*
Noise of construction	3.9	2.4*	11	13
Workers coming into the area	4.8	3.7*	52	22*
Noise of station in operation	4.3	3.2*	13	12
General appearance of power station	3.9	2.2*	32	48*
Area of land fenced off	4.0	2.3*	19	26*
Steam from station when operation	4.2	2.6*	7	21*
Increased security and policing	5.2	3.7*	15	11
Warming of sea water	5.0	3.7*	14	12
Transportation of nuclear waste	3.1	1.9*	49	55
Overhead power cables pylons	3.0	2.0*	40	37
Overall height of buildings	3.9	2.1*	15	38*

[a] Possible range of scores from 1 (very much for the worse) to 9 (very much for the better).

[b] The scores represent the percentage of subjects selecting each factor among the five most important. The columns do no add up to 500 because of the inclusion of subjects who chose fewer than 5 aspects.

* Differences between groups significant at $p < 0.05$.

Source: Adapted from van der Pligt, Eiser and Spears (1986b), p. 82.

among the five most important by the two groups. The results show that, irrespective of familiarity, respondents attach great importance to the issue of transportation of nuclear waste. More than 50 per cent of the present sample selected that issue as among the five most important.

Table 4.7 Perception of consequences of a local nuclear power station as a function of familiarity

	Mean score[a]		Importance[b]	
	High familiarity (N = 218)	Low familiarity (N = 430)	High familiarity (N = 218)	Low familiarity (N = 430)
Employment opportunities	8.2	6.6*	61	31*
Tidiness of the village	4.4	3.6*	8	13
Standard of local recreational facilities	5.3	4.9*	7	8
Social life in the neighbourhood	5.1	4.7*	17	10*
Wild life	3.6	2.2*	33	57*
Marine environment	4.1	3.0*	23	28
Farming industry	3.8	2.5*	30	45*
Security of local electricity supplies	5.5	5.6	19	9*
Health of local inhabitants	4.2	3.2	36	40
Landscape	3.4	2.0*	31	53*
Holiday trade	4.8	3.2*	7	18*
Business investment	6.3	5.0*	22	15*
Personal peace of mind	3.9	2.7*	33	35
Standard of local transport and social services	5.5	5.5	20	10*
Standard of shopping facilities	5.4	5.4	17	1*

[a] Possible range of scores from 1 (very much for the worse) to 9 (very much for the better).

[b] The scores represent the percentage of subjects selecting each factor among the five most important. The columns do no add up to 500 because of the inclusion of subjects who chose fewer than 5 aspects.

* Differences between groups significant at $p < 0.05$.

Source: Adapted from van der Pligt, Eiser and Spears (1986b), p. 85.

Further analysis revealed that respondents familiar with living near a nuclear power station differed most significantly from the remaining respondents in terms of the importance attached to 'workers coming into the area'.

Table 4.7 shows the mean ratings of the fifteen (mainly long-term) effects of the building and operation of a nuclear power station. Results confirm those shown in table 4.6 and indicate a more pessimistic view of the various potential outcomes by non-familiar respondents. The most striking differences concern the effects on employment opportunities and the adverse consequences for the environment and public health. Finally, non-familiar respondents predicted a more negative impact on their peace of mind than did those living near the existing nuclear power stations. Both groups, however, thought their 'peace of mind' would be negatively affected by the building and operation of a new nuclear power station. Closer inspection of the overall mean scores reveals that both groups, on average, think that most factors will have a negative impact on life in their community. Both table 4.6 and table 4.7 show a majority of mean scores lower that 5 (midpoint of the scale).

We again conducted a number of analyses of covariance with familiarity as an independent variable and the attitude-index score as a covariate. Results of the these analyses revealed that attitude had a more pronounced effect than familiarity. Most items, however, still showed significant effects due to familiarity. (For a more detailed account of these analyses seen van der Pligt, Eiser and Spears, 1986b.)

Table 4.7 also shows which items were chosen as the five most important by the two groups of respondents. The results show substantial differences between the two groups (familiar versus non-familiar) on a number of items. Again, we conducted further analyses to determine which aspects most distinguished the two groups. The results revealed four items that had considerable predictive power in separating the two groups. The item which was most different from the two groups concerned 'employment opportunities' (seen as far more important by familiar respondents). Further significant contributions were made by three items seen as more important by nonfamiliar respondents ('wild life', 'holiday trade' and 'landscape').

Closer inspection of tables 4.6 and 4.7 indicates that the inclusion of both the perception of the various potential consequences of a nuclear power station and the perceived importance attached to each of these consequences provides a more complete picture of the attitudinal differences between the two groups. Results shown in table 4.6 suggest that non-familiar subjects are generally more pessimistic about workers coming into the area and the appearance of the buildings. They also find the latter aspect more important than do 'familiar' respondents at Hinkley Point. Table 4.7 shows that non-familiar respondents see the various risks of nuclear power stations as both more serious and more important than do the Hinkley Point sample. This group, on the other hand, is more optimistic and attaches greater value to the economic benefits.

Some cautionary notes seem in order. For instance, Thomas and Baillie (1982) suggest that alternative explanations of the effects of familiarity should not be totally discarded. They argue that communities initially more favourable may have been chosen in the first place; alternatively, those antagonistic towards the proposals might have moved away or did not move into the area. They caution against defining familiarity purely in terms of spatial distance and suggest that economic motives may also be important. They quote from a study conducted around British Nuclear Fuels' Sellafield plant, where employees had more positive attitudes than non-employees towards the facility.

Our own findings indicate that low familiarity is associated with greater pessimism about both long- and short-term effects, greater involvement and stronger anti-nuclear views. However, the fact that there was no power station in the locality of the low familiarity group does not necessarily mean that residents were unfamiliar with nuclear power. There had been vigorous local (and national) press coverage of the issue some eighteen months before the data collection. In other words, the findings may be more a reflection of adverse public reactions to an unwanted proposal. This alternative explanation is not supported by our covariation analysis but some authors point to the need to define familiarity not only in terms of geographical (spatial) distance but also in terms of social distance (e.g., how many people are known who work at a power station).

For instance, Lindell and Earle (1983) described spatial distance as a determinant of perceived risk, but in combination with an experiential factor. Their findings showed that those who were prepared to live or work closest to a nuclear facility were nuclear engineers, followed by chemical engineers, science writers, and members of the general public having experience of other hazardous facilities. However, only 14 per cent of the general public were prepared to work within 16 km of a nuclear power station.

Conclusions

This chapter has focused on local attitudes towards nuclear power stations. Not surprisingly, these attitudes are more extreme and more negative than attitudes towards nuclear energy in general and/or the building of nuclear power stations elsewhere. As was argued before, 'not in my backyard' seems to be the generally favoured location of nuclear power stations. This phenomenon is not restricted to nuclear power stations but also applies to other potentially hazardous facilities such as chemical plants.

A limited number of studies have focused on the development of local attitudes during construction and siting procedures. Results suggest a general perception of (local) benefits being outweighed by the (local) costs. Attitudes tend to be quite extreme and most local residents see both immediate and long-term negative consequences. Attitudes in localities around existing nuclear power stations tend to be slightly less unfavourable. Policy developments in the late 1980, suggest that the nuclear industry and the relevant authorities prefer to locate nuclear power stations and nuclear-waste facilities in 'nuclear parks'. This more accepting attitude in localities already familiar with nuclear facilities is, however, quite unstable and fragile. Acceptance of inequity (i.e., local costs versus national benefits) has its limits. It could well be that even localities already hosting nuclear facilities will be less and less prepared to accept further expansion of the nuclear industry in their locality. To this and related issues I turn in the next chapter in which the siting of nuclear-waste facilities will be discussed.

5

Siting nuclear waste facilities

During the 1980s the need to dispose of nuclear waste resulted in considerable conflict. In several countries decisions about waste disposal sites were frequently postponed because of the sensitivity of the issue. For instance, in the UK just before the 1987 general election the Conservative government announced a postponement of its decision to select a site for intermediate-level radioactive wastes. In the late 1980s the US Department of Energy postponed consideration of sites for the country's second high-level radioactive waste repository. Several other countries (such as Germany and the Netherlands) also decided to postpone consideration of sites for high-level waste in the wake of extensive research into the suitability of alternative storage methods. In the UK the Nuclear Industry Radioactive Waste Executive (NIREX) carried out a site-selection procedure which resulted in the announcement that future investigations would be limited to two sites, Dounreay and Sellafield. Both already hosted nuclear power facilities. Planning applications for further investigation at Sellafield were expected to be made towards the end of 1992 (Tasker, 1989a). The late 1980s also saw the postponement of deadlines for several states in the USA to open their own low-level radioactive waste facilities – a result of the reluctance to impose these facilities bluntly on resistant local communities. So-called 'back-end' wastes (i.e., wastes produced at the end of the nuclear power generating process) have accumulated in all countries that operate reactors. Hare and Aikin (1984) report a total of 26 nations having serious problems with nuclear waste. 'Not in my back yard' (NIMBY) seems to be the generally favoured location for nuclear wastes.

In this chapter I will first present a brief history of the technological and policy context of waste disposal; next I will discuss possible sources of public concern. This will be followed by a summary of research on the psychological effects of hazardous waste facilities and on ways to improve the choice of sites for these facilities.

Technological and political context

Nuclear power generation includes a large number of processes between the mining of uranium and the ultimate disposal of radioactive wastes. Once uranium ore has been mined and enriched, it is turned into fuel to be used in nuclear reactors. In the reactor, energy is generated by splitting the U-235 atoms contained in the fuel rods (nuclear fission). This process in fact creates two main products: first, an enormous amount of heat which is used to generate electricity; secondly, fission products, which are lighter nuclei formed by the splitting of the original nuclei. Most of these fission products are intensely radioactive. Reactor operation also creates many different radioactive substances apart from the lighter nuclei. Part of the uranium is converted into other heavy nuclides, of which plutonium-239 is the prime example. Others include americum, curium and other isotopes of plutonium. Still other radioactive wastes are formed from the materials of which the reactor is built (e.g., cobalt, iron). Each of these radioactive substances has its specific characteristics and (unalterable) rate of decay. When the number of fissionable U-235 atoms in the rods diminishes below a certain level, the rods – constituting nuclear waste – are removed. The spent rods then go into temporary storage in large ponds of water for a 'cooling-off' period of at least five months – allowing short-lived radioactive byproducts to decay.

From these holding ponds, spent fuel rods can follow one of two routes towards final 'disposal'. They can be transported directly to a permanent waste-management facility, designed to isolate the spent fuel for thousands of years. Another option is to ship the rods to a reprocessing plant, where the unconsumed U-235 and the fissionable plutonium can be extracted and manufactured into new fuel rods. The wastes left after reprocessing are then solidified and transported to permanent waste storage facilities. Again, these facilities should be designed to isolate the waste for thousands of years.

In 1992 this 'back end' of the fuel cycle remained inoperative. For instance, nearly all of the 385 million litres of high-level radioactive wastes produced by US military and commercial reactors were held in temporary storage, either as spent fuel rods kept in the holding ponds at each of the operating commercial reactors or in the steel tanks at military installations and reprocessing plants. Similarly, the bulk of the backlog of high-level waste in Britain still needed to be vitrified and was stored temporarily in stainless steel containers at the Sellafield site. Such bottlenecks in the fuel cycle were apparent in most countries and resulted at least partly from indecision on whether to promote commercial development of the breeder reactor which would use reprocessed fuel. France was already engaged in a fast-breeder programme and the UK had decided to collaborate with the French in order to develop a new fast-breeder reactor. Overall, the situation with respect to high-level waste was similar in most Western countries. Most waste was stored in temporary facilities.

The most significant obstacle to the high-level radio active waste problem concerns the difficulty of finding a method for permanently managing wastes which is both technically feasible and politically acceptable. In the early 1990s political and scientific uncertainty prevailed, while spent fuel rods accumulated in cooling ponds at many nuclear reactor sites. As temporary storage facilities filled up, the decision about what to do with the wastes became more critical. Exporting the wastes was not a viable option either. The political climate more or less excluded the possibility of exporting wastes to the developing world.

Bearing in mind that radioactive byproducts are extremely dangerous, resolving this issue seems essential. For instance, iodine-131 produces cancer of the thyroid gland when absorbed internally. Strontium-90 can enter the food chain and cause bones to disintegrate. Plutonium is perhaps the strongest carcinogen of all. When inhaled, even in particulate form, it is lethal; when it enters the body through wounds, plutonium can produce cancers.

Around 1960 scientists in the USA first accepted the necessity of developing safe methods of storing nuclear wastes. The wastes generated up to that point were under military control and were stored in large carbon-steel tanks, but they required eventual transfer to a more permanent form of isolation. Storage was also extremely expensive because a large volume of wastes with only a low concentration of radioactive elements had to be stored in leakproof containers. In 1957 and 1958 two major scientific appraisals of the waste disposal problem were presented to the US Atomic Energy Commission (AEC). These studies indicated that the problem of atomic waste disposal would, to a large extent, determine the future of nuclear power development in the private sector.

After these early appraisals were made a variety of alternative storage schemes were suggested. The most popular proposal involved depositing wastes in large-scale stable underground geological features such as abandoned salt mines (Weinberg, 1972). Salt is favoured because of its low porosity and permeability. Moreover, it has natural convergence proportions which seal any faults that could provide a pathway for the release of radionuclides. Disposal in clay has also been considered, mainly for its powerful retention capability and the deposit thickness. Granite formations are another option, mainly because granite is inalterable, with high mechanical and chemical resistance. Other solutions, such as injection down deep wells, dumping casks of wastes into deep and relatively immobile bodies of water in the ocean, dumping glass containers on the bottom of the Antarctic icecap, and surface storage in individual concrete mausolea have all been met by severe criticism or painful facts (e.g., the detection of traces of plutonium and cesium on the sea floor near Pacific burial sites used in the early 1960s).

On the basis of this experience Fallows (1981) argues that the debate on nuclear waste illustrates the consequences of the low level of agreement within the scientific community and among government regulators. This inability of experts to agree on safe and acceptable solutions has resulted in public confusion and debate and a stalemate on where and how to dispose of nuclear wastes. It seems that even after more than three decades of nuclear energy, the failure to put into operation an adequate 'back end' of the nuclear fuel cycle continues to haunt the prospects for one of the world's major energy sources. For instance, in the USA spent fuel is accumulating at the eighty-two operating commercial reactors, while on-site temporary storage space at some reactors will be filled in a matter of several years.

In various countries low-level radioactive waste siting programmes, which require only a relatively small number of facilities, have also run into major problems. Here the social issues concerned are broader, since waste generators include industries, hospitals, and biomedical research facilities, as well as nuclear power plants (see, e.g., Decima, 1987; Welch, 1985).

Public concern and opposition

When, in the early 1960s, the New York State Atomic Development Authority announced plans to build an atomic service centre in western New York, residents in the region supported the project as an attractive opportunity for the depressed local economy. The centre was constructed and began operations in 1966. It was not until the mid 1970s that local groups began to question the safety of the facility's operations and the wisdom of turning their region into a nuclear dump (cf. Fallows, 1981).

In 1970 two proposals to create waste management facilities led to vigorous local protests. An attempt to locate a dump for intermediate-level radioactive wastes in a sparsely inhabited portion of rural Oregon was abandoned after protests by local residents who feared the development would jeopardize their water supply and property values. In the same year plans were announced to utilize an abandoned salt mine in a sparsely populated region in Kansas, as the USA's first permanent, underground nuclear waste repository. The Kansas State Geological Survey criticized the technical integrity of this proposal and by 1971 the plans had been shelved as a result of the technical criticisms and the subsequent political pressure (cf. Fallows, 1981).

In addition to concern about these specific projects, the general public became more interested in the issue largely as a consequence of conservation groups ventilating their concern about the possible negative environmental effects of nuclear facilities. By 1973 the issue of waste disposal was becoming increasingly central to policy discussions about nuclear energy in general. For example, the American champion of consumers' interests Ralph Nader described the unresolved questions of transporting and disposing of radioactive wastes as the two major dangers of operating nuclear reactors (*NYT*, 1973).

During the mid 1970s public concern with nuclear power, and with waste disposal in general, increased dramatically. A variety of opinion polls indicated that the public saw nuclear waste as the most important unresolved issue of the large scale development of nuclear energy (see also Chapter 1).

In California enough voter support was gathered throughout the state to place a proposition on a ballot (Fallows, 1981). This proposition would have required utility applicants to prove to the public the safety of any radioactive waste disposal facility before they could receive an operating licence from the state. Similar campaigns were initiated in seventeen other states in the USA. More and more concerned interest groups emerged, frequently supported by 'defecting' scientists and engineers who had left their jobs, to protest against the further development of nuclear power. In Western Europe local authorities proclaimed 'nuclear free zones' in order to block both the transport of nuclear wastes through their area and the possible siting of nuclear waste facilities.

Protests over specific proposals to site waste disposal facilities became more vehement. Attempts to begin exploratory drilling were interrupted, and sometimes had to be abandoned, due to local protest. From the mid 1980s onwards attempts to control the development of nuclear power in general and nuclear waste facilities in particular came from local and regional (state, county or province) legislative bodies which passed resolutions superseding the central government's jurisdiction in the siting and licensing of nuclear waste storage facilities. In other words, quite often attempts to find locations for nuclear waste facilities tended to result in a conflict between, on the one hand, local residents and local government and, on the other hand, central or federal government and the relevant industry.

Siting efforts thus encountered determined, and often vehement, local opposition. The US Environmental Protection Agency remarked that the issue seemed to unite 'grandmothers and US Congressmen, factory workers and university scientists, those who never graduated from high schools and those with doctorates in ecology and physical sciences' (USEPA, 1979). The vehemence of the opposition often stunned developers and regulating agencies. Kasperson (1985) notes an example where angry citizens were prepared to blow up a facility and there were reports of threats of death or physical harm to key individuals and their families (USEPA, 1979, iii). In the face of such volatile local response, it is not surprising that the US siting record is rather bleak: approximately 32 siting efforts occurred between 1979 and Sepbember 1984; 5 treatment and 2 storage facilities were *approved*. Only one waste disposal facility was successfully sited and constructed, and that (in Maryland) was subsequently closed because it could not compete economically with previously sited facilities (Ryan, 1984). A conference on radioactive waste management in London in 1989 concluded that the number of sites available for the disposal or storage of nuclear wastes is limited to those places where the local public has already accepted the nuclear industry (Emmings, 1989). Moreover, the indications in Europe are that this situation is unlikely to change for quite some time.

This brief overview indicates that public anxiety over nuclear waste management has grown into a major policy issue. Both locally affected groups, local governments and national environmental and political organizations have joined the opposition. The early political discussions tended to focus on technical problems of alternative solutions to nuclear waste disposal. Such concerns remained, but the dispute moved beyond the scientific community. Continuing technical uncertainty led to questions about the ability of the relevant authorities to act in the public interest and the legitimacy of allowing them to do so.

The question is whether ultimately a waste management system can be decised and implemented that is acceptable to a wary and distrustful public. The importance of public acceptance is now accepted by most involved parties. They have been convinced by the experience of desperate and turbulent searches for storage sites for toxic waste in the

United States, the Federal Republic of Germany, the Netherlands and the UK. Some nuclear experts have singled out public acceptance as the most formidable obstacle confronting nuclear power (Brooks, 1976; Weinberg, 1977). Belatedly, the reality has become official policy. Kasperson (1980) illustrates this point with the following quotes:

> A US Nuclear Regulatory Commission task force concluded in 1978 that 'past failures of proposed radioactive waste-management systems have stemmed in large part from neglect of nontechnological necessities in the implementation of systems' (Bishop, 1978, p. 57).

> The 1979 Interagency Review Group on Nuclear Waste Management argued in its *Report to the President* that: 'the resolution of institutional issues... is equally as important as the resolution of outstanding technical issues and problems' and that such resolutions 'may well be more difficult than finding solutions to remaining technical problems' (Interagency Review Group on Nuclear Waste Management, 1979, p. 87).

Recent statements from the UK confirm this viewpoint. Eyre and Flowers (1991) state that 'radioactive waste management, alongside reactor safety and electricity costs, is one of the important areas where the nuclear industry has failed to communicate the facts'. Moreover, the 1976 Royal Commission Report (Royal Commission on Environmental Pollution, 1976) also acknowledged that it has proved very difficult to regain public confidence. This seems particularly evident when it comes to siting nuclear waste repositories. Yet, as Kasperson (1980) notes, to recognize that a problem exists 'out there' is not to understand or to act upon it. In 1992 waste programme deliberations reflected what was perhaps the first stage of corrective action; namely, that the problem was recognized to be something different from what it had been before.

Attitudes, risk perception and equity

As the previous section suggests, local attitudes towards nuclear waste facilities tend to be negative. For instance, Mazur and Conant (1978) studied public attitudes towards a proposed long-term nuclear waste storage facility near Syracuse, New York. They interviewed local residents during a brief flurry of publicity about the proposal and again nearly four months later, after the publicity had died down, to assess the stability of attitudes.

Near Syracuse lies a large salt formation which the United States

Energy Research and Development Administration (USERDA) intended to investigate to assess its suitability for long-term storage of nuclear waste. Local politicians immediately protested. The two major Syracuse newspapers picked up the story and supported the local view. Several stories appeared in newspapers and on television over the next few weeks. The media coverage was neutral to negative towards the project; Mazur and Conant (1978) could not find any favourable comment. Within four weeks the publicity ended and the controversy appeared to be over, with no apparent effect on ERDA's plans.

Those who had heard about the issue (63 per cent of the men, 21 per cent of the women) strongly opposed the plans; only 20 per cent of these respondents favoured the placement of a nuclear waste storage facility near Syracuse. Of the people who had not heard about it, 29 per cent were undecided and 26 per cent were in favour. Exposure to the controversy made women's attitudes especially more negative. In all, the vast majority opposed the plans and this attitude persisted over time.

Other studies resulted in similar conclusions. Brooks (1976, p. 52), for instance, noted that 'almost from the first consideration of nuclear power as a realistic possibility, the public has viewed radioactive waste management as the primary obstacle to its ultimate technical success and social acceptability'. A variety of empirical studies underlines this sombre appraisal. A 1976 Harris Poll revealed that radioactive waste disposal topped the list of major public concerns over nuclear energy, with some 65 per cent of the public citing it (Harris and Associates, 1976). A study made in 1976 of voters' attitudes in Sacramento County towards the 1976 California initiative concluded that 'the principal drawback of nuclear energy plants in the perception of our respondents was specifically the difficulty of safe waste disposal...' (Groth and Schutz, 1976, p. 15). A detailed analysis of all relevant opinion surveys conducted for the US Department of Energy found evidence that wastes were at or near the top of concerns of nuclear energy (Nealey, Melber and Rankin, 1983; see also Chapter 1). Opinions polls in most Western countries (with the possible exception of France) indicate an extremely low acceptance of nuclear wastes storage facilities.

Initial attempts to explain these reactions focused on the 'irrational fear' of the public. References have been made to 'panic' and 'irrationality' with regard to radiation and chemicals (e.g., Cohen and Lee, 1979; Dupont, 1980; Bord, Ponzurick and Witzig, 1985). Pahner (1976) has suggested that public concern over nuclear risks stems from pre-existing images of the horror of nuclear war, fears related to the invisibility of radiation, uncertainty about exposure, and fears of the immediate and delayed effects of radiation on genetic processes. Dupont

(1980) argued that many people have a 'phobia' about nuclear power. (See Chapter 1 for a more detailed discussion of this view.)

The presence of fear in public reactions towards nuclear waste is indisputable. It is disputable, however, that fear is the dominant factor. No explicit studies of fear in attitudes towards the wastes themselves have been conducted. Studies of the fear hypothesis for nuclear power have not supported the proposed relationship (Cunningham, 1985; Freudenburg and Baxter, 1984). Brown, Henderson and Fielding (1983) found concern (a cognitive concept) about nuclear power to be far more pervasive than anxiety (emotion). As we saw in Chapter 1, a model of nuclear power perceptions which included a 'phobia' submodel – consisting of fear of explosions, ignorance about nuclear energy, and lack of education – explained between 5 and 10 per cent of the variance (Mitchell, 1984). Other factors (largely ideological) made a more significant contribution to the explanation of people's perceptions of nuclear power issues.

Johnson (1987) notes that despite the lack of evidence for irrational fear as the primary factor in attitudes towards nuclear waste, it should not be dismissed entirely. Fear of personal or family exposure to radiation risks is more salient for local residents in the context of siting waste facilities. It is clear, however, that dismissing public concerns as entirely irrational is not likely to enhance the acceptability of waste facilities. Instead, more attention should be paid to risk–benefit equity and safety.

As was argued before (Chapter 2), nuclear power carries risks that pose specific difficulties for both the expert and the lay public. The risks concern low probability, but highly catastrophic consequences, which, given the short history of the technology, require that modelling and simulation substitute for experience. Depending upon the degree of confidence in risk-assessment techniques, the valuation of the future, and the expected changes in the future of the technology, scientists can in good faith differ greatly in their assessment of the risk. Moreover, new problems such as theft, human error and sabotage continue to surface and need to be taken into account. Given this disagreement among experts and the nature of the risks it is not surprising that the public debate tends to be dominated by safety issues.

A survey in Massachusetts in 1983 of attitudes towards a hazardous waste facility in each of five communities underlines the importance of these risk and safety issues (Portney, 1983). Although the survey found considerable differences between the five sites, opposition was clearly linked to concerns about safety. Moreover, in testing some eleven different proposals designed to move people from opposition to support, most opponents indicated that they would not change their minds. Among the minority indicating willingness to change on certain conditions, greater assurance about safety and more safeguards were clearly most influential. A variety of financial incentives, by comparison, had considerably less impact. Finally, opposition did not appear to be rooted in lack of information as the opponents tended to be somewhat more knowledgeable than the supporters.

Perceived risk and more general values in public acceptability of nuclear energy (see Chapters 2 and 3) apply equally to the nuclear waste issue. In other words, it seems that perceived risk is a central problem in hazardous waste facility siting and that public perceptions are rooted in 'objective' characteristics of these risks. It is also the case, however, that the risk issues interact with related value conflicts, and that the underlying attitudes are likely to be persistent and difficult to change.

Siting processes, however, have another major characteristic which is possibly of even more importance than the safety aspect: inequity. Inequity, it is widely held, is a crucial problem for hazardous waste facility siting. Indeed, for many it is *the* problem. This is illustrated by the following from the 1981 policy statement of the US National Governors' Association:

> Once a site is identified, the community objects to being the dumping ground for the state or region. It opposes the proposed facility because the benefits will flow to the owner, operator, waste-generating industries, and the public at-large (which fears 'midnight' dumping), while the *risks* will be concentrated locally – in their community (US National Governors' Association, 1981).

It has been argued for some time that facilities providing diffused benefits while imposing high costs at or near the specific location are bound to engender opposition (see, e.g., Kasperson, 1985). Equating the source and ultimate repository of the waste as much as possible – either on a community or regional basis – could prove a more fruitful approach. Generally, an acceptable balance of risks and benefits will be very difficult to obtain. However, some progress has been made on the issue of siting toxic waste facilities.

Improving equity

Equity means different things for different people. Recent usage in the social sciences takes three major approaches: *utilitarian, distributive,* and *procedural* (Enbar, 1983). Most of the discussion of siting equity, however, has focused upon distributive issues, as was suggested in the US National Governors' Association statement noted above.

Such equity statements consist of two major components (see also Kasperson, 1985). The first component is a statement of the distribution of benefits and harms affecting a specified population which would result from a given decision (i.e., a siting decision). This requires an analysis that includes (a) a specification of the economic, social, psychological and health-related aspects whose distribution is being investigated; (b) an explicit delineation of the population and relevant subpopulations to be considered in the analysis; and (c) a statement of the actual impact distribution as defined by (a) and (b) – which would result from siting a hazardous facility in a particular location.

The second component is a set of normative standards or principles by which the equity or 'fairness' of particular distributions may be judged and by which the social preferability of one distribution over another may be determined (Kasperson, Derr and Kates, 1983, p. 332).

Unfortunately, the current empirical understanding of impact distributions that would result from the siting of a hazardous waste facility at a particular site is quite limited. Reasons for this limited knowledge include: (a) the relatively underdeveloped state of theory supporting approaches to social impact assessment; (b) the limited experience of siting facilities in recent years; (c) the highly site-specific nature of social impacts (cf. Kasperson, 1985). In one of the few elaborate analyses of equity at a hazardous waste site, Kates and Braine (1983) painted a complex picture of gains and losses over more than a dozen locations stretching across the entire United States: benefits for some corporations, institutions or governments and local residents; losses for others; a mixed balance for yet others. One of the major problems of this type of impact assessment is the limited knowledge about socio-economic impacts and delayed health effects, and how to translate these possible futures into immediate compensatory (economic) benefits.

These first attempts to find equitable solutions for the siting of toxic waste facilities indicate that acceptable solutions for radioactive wastes involve a series of important value issues. Because of the limited experience it is not clear what all the relevant issues are. Yet it is clear from public reactions that certain question are pervasive: *who pays?* (with either money or health); *who benefits?* and *who enjoys what rights?*

Kasperson (1980) identified three control themes from the array of ethical concerns which must be faced. First there is the 'locus problem'. The storage of radioactive wastes at particular sites involves diffuse benefits to society as a whole. For the local populations, however, there are concentrated risks (e.g., the possibility of accidents, chronic leakage, purposeful intrusion) as well as some benefits (e.g., employment, taxes). Secondly there is the *'legacy problem'*. The benefits of nuclear energy from particular power stations are concentrated in several decades yet radioactive wastes remain dangerous for thousands of years. The question arises as to whether risks and burdens should be exported to future generations. Thirdly there is the *'labour/laity problem'*. A waste-management system may be designed to maximize either public health or worker safety, but sometimes not both simultaneously (at least not at feasible cost). Under what conditions should society knowingly increase the risk to the worker?

Most attention has been given to the first two problems. In 1980 Kasperson concluded that the relevant agencies have generally failed to develop these equity principles. In the early 1980s prevailing approaches tended to be piecemeal and internally inconsistent. However, since then a number of attempts have been made to apply equity principles to siting processes.

One way to resolve the risk-benefit inequity is to keep radioactive wastes at their place of origin. 'On-site' disposal has generally been considered less disruptive to local residents than 'off-site' disposal. First, these wastes are 'familiar' to local residents near a nuclear power station. Secondly, the waste disposal site by an existing facility generates extra local economic benefits. This option has seemed to gain in popularity. Proposals for waste facilities in the Netherlands and the UK in the early 1990s focused on existing nuclear sites as sites for waste repositories. Several studies have suggested that experience is responsible for the lower perception of risk from nuclear power plants and other hazardous facilities. Usually these studies have compared people who live close to a facility or plant with people living at some distance (see, e.g., Earle and Lindell, 1982; Ester et al., 1983; Morell and Magorian, 1982; Rogers, 1984; Dew and Bromet, 1985).

Up to 1992 there had been no record of local concern about risks from spent fuel rods kept in holding ponds at nuclear power plant sites. Some have argued that this presumed positive view may be caused by the need to reduce cognitive dissonance as a result of working or living in a place perceived to be dangerous (e.g., Ester et al., 1983; Dew and Bromet, 1985), but no direct evidence for cognitive dissonance has been found (Rogers, 1984). Lyons, Freeman and Fitzgerald (1986) found that over 70 per cent of respondents in a survey supported the idea of wastes being stored on the site where they had been created. Only a third approved of a lottery procedure by which all communities would have an equal chance of being selected as a site of waste disposal.

Unfortunately this research did not deal with a specific site but just raised the general issue. As Johnson (1987a) remarks, the above 70 per cent could all be thinking of facilities not in their own locality. The familiarity criterion is implicit in many decisions on siting nuclear waste. For instance, the US Department of Energy's selection of final candidates for the USA's first high-level radioactive waste repository consisted of locations whose populations had extensive experience of living with nuclear facilities. Similarly, the Dutch government selected a locality with an existing nuclear power plant (Borsele) as the first site for low- and intermediate-level radioactive waste. Finally, the British government has opted more and more for the creation of 'nuclear parks'; i.e., the location of several nuclear facilities at a single site (e.g., Sizewell, Hinkley Point).

Yet events from the late 1970s onwards have questioned the positive effects of proximity and familiarity. Host community support for nuclear power plants has apparently decreased sharply since the accident at Three Mile Island in 1979 (Freudenburg and Baxter, 1984). Our own research in Southwest England has also indicated considerable public opposition in host communities (van der Pligt, Eiser and Spears, 1986b, 1987b). Similar findings were obtained near Sizewell (see also Chapter 1). Johnson (1987a) reports concern in the Hawford Nuclear Reservation, a location with 42 years of experience with radioactive materials. In general, familiarity or experience of nuclear power facilities can improve local acceptance of waste repositories but it will not necessarily generate local approval.

If familiarity alone is an insufficient precondition for local acceptability, the economic benefits of a waste disposal facility could possibly offset the perceived risks. There is some evidence suggesting that direct economic benefits increase local acceptance. For instance, members of local citizens groups in favour of restarting the TMI Station's undamaged second reactor after the 1979 accident were generally in a position to benefit economically from the restart (e.g., shopkeepers); opponents were not (Soderstrom et al., 1984). Similarly, the absence of potential benefits for locality may increase local opposition. A study of local perceptions of the West Valley nuclear reprocessing plant in the USA illustrates this point (see Kates and Braine, 1983). As residents realized that the promised benefits of the facility were uncertain and that other economic developments (e.g., tourism) might be harmed by the growing controversy, local concern increased significantly.

In general, however, the direct local economic benefits of waste disposal facilities are relatively limited. Construction activities bring in many people from elsewhere and disrupt local life, while the number of people that are employed when the facility is operational is quite modest and the number of locals who obtain employment at a facility is possibly even lower. Not surprisingly, a low-level radioactive waste facility proposed for Pennsylvania was opposed even though the local economy was declining (Bord, Ponzurick and Witzig, 1985).

It seems therefore that the combined effects of familiarity and economic benefits do have some impact but fail to guarantee local acceptance. However concentration of waste disposal as close as possible to the origin of the waste has a number of drawbacks. Putting wastes (e.g., low-level radioactive wastes) in urban, industrial areas on the ground that these areas get most of the direct benefits from the generating industries is usually inadvisable because of health risks. The use of rural sites remote from population centres reduces such risks. Geological factors may also prohibit on-site waste disposal. Morell and Magorian (1982) argue against on-site disposal as a general rule. They state that the limited direct benefits are generally insufficient to compensate for the perceived risks. Apart from this, on-site disposal is unlikely to be sufficient to keep wastes where they are generated, given the quantities of waste being produced.

The off-site disposal of radioactive waste creates a hazardous situation which can be seen as the 'import' of problems that have been created elsewhere – a typical case of concentrated costs and diffuse benefits. Generally, localities will consider the imposition of a waste facility as inequitable and will tend to oppose it. Several authors describe local unacceptability of chemical wastes (e.g., Morell and Magorian, 1982; Burns, 1984; Fitchen, Heath and Fessenden-Raden, 1987). Johnson (1987a) argues that the 'foreignness' of the waste is even more salient in the case of nuclear waste not stored at nuclear power plants. In devising ways to reach more equitable means of locating off-site waste disposal facilities several aspects are important. Four of these will be discussed in the remainder of this chapter: (1) economic compensation; (2) safety and control; (3) trust and risk communication; (4) public participation in decision-making.

Economic compensation In the USA economic compensation is used as a major part of radioactive and chemical waste disposal policy in order to increase the acceptance of toxic waste disposal facilities. Most experience with this approach has been obtained from the location of chemical waste sites. It indicates that the economic approach to the siting of waste facilities – if handled with care – may reduce, but not eliminate, local opposition.

Some studies found more favourable attitudes were produced by such incentives as the prospect of new jobs, lower taxes and the protection of property values (Lyons, Freeman and Fitzgerald, 1986; Fowlkes and Miller, 1987). In Pennsylvania the planned reduction of nuisance during the construction phase (by building a special access road) and protection of house values had a positive impact on the acceptance of a proposed low-level radioactive waste facility (Bord, 1987).

The main difficulties of this approach are equating health, nuisance and money. Some have argued that it is impossible to equate reduced health or life expectancy with a specific sum of money (e.g., Foa and Foa, 1980). Others argue that compensation for nuisance, property values and health risks caused by the transportation of waste to a disposal site are difficult to translate into compensation payments, because of the tremendous variations within a community due to the different risk levels and/or distances to the route to a disposal site. In other words, it is difficult to make equitable compensation on a geographical basis. Johnson (1987a) concludes that carefully planned compensation packages can reduce some opposition to local waste facilities. This optimism seems to apply primarily to chemical wastes and, possibly, low-level and intermediate-level radioactive wastes.

Plans for the disposal of high-level radioactive waste attract vehement opposition, indicating that the redress of risk–benefit inequities cannot be achieved only by economic methods. Some even argue that the provision of financial compensation could be seen as a 'bribe', which would only enhance distrust and suspicion of the relevant authorities.

Direct and indirect financial incentives have been successfully employed in France, where direct payments are given to host communities of nuclear waste storage facilities (Starr, 1980; Carle, 1981). Because perceived risk so dominates local public response, risk minimization, risk allocation, and risk-sharing questions are the dominant issues in equity problems.

Safety and control Kasperson (1986) notes that when dealing with the distribution of risks and benefits there is little prospect of restoring a community's original conditions (i.e., the situation before the introduction of the hazard) through benefits. For most residents safety is the crucial issue. Swartzman, Croke and Swibel (1985) investigated the role of compensation and safety. Their study attempted to determine whether public opposition to the siting of hazardous waste facilities might be reduced by proposed programmes of compensation or enhanced environmental monitoring. They found that both reduced public opposition. Bachrach and Zautra (1986) found that respondents generally attached greater importance to safety issues (such as frequent monitoring, improved medical facilities and emergency systems) than to economic compensation.

Carnes et al. (1982) made a direct comparison of economic and safety aspects as incentives in helping resolve siting conflicts. They found that approximately 20 per cent of their sample (420 rural Wisconsin residents) abandoned their stated opposition towards the siting of a nuclear waste repository. Their findings indicate that non-monetary incentives, such as independent monitoring and access to information, may significantly contribute to public acceptance of radioactive waste facilities. Moreover, control seemed to be a key factor. Several authors have pointed to the need for an independent research institute or agency to evaluate the risks of waste disposal (Kasperson, 1986; Black, 1987; Johnson, 1987a). Increased monitoring and control are both assumed to be important. The desire for increased monitoring is frequently mentioned in survey research. It has been proposed that people could be given simple manuals to help them assess possible health effects and also become more knowledgeable about them (Johnson, 1987a). Increased control implies sharing power, which tends to be complicated and time-consuming. Bord (1987) found that in Pennsylvania the best way to obtain support for a low-level nuclear waste facility was to grant citizens the right to shut it down. Furthermore, shared power options were generally far more popular than mitigation and compensation. Notwithstanding the time needed to implement shared-power solutions, it seems worthwhile to investigate this possibility further given the above findings.

Trust and risk communication As was indicated in previous chapters, the nuclear industry has a history that fosters distrust. Too often particular risks have been mismanaged, neglected or denied. This has resulted in public scepticism and distrust. For instance, a survey of Wisconsin residents revealed that most respondents believed that the government was moving too slowly to solve the waste disposal problem or that it was uninterested in the opinions of local citizens (Kelly, 1980). Similarly, a survey of attitudes towards the siting of hazardous waste facilities revealed that many respondents would not trust the management of companies that operated treatment facilities and the government regulators who oversaw them (Portney, 1983). Johnson (1987a) suggests that there are few ways of overcoming such distrust. Possibilities include the acknowledgement of past errors (confession and penitence); the development of programmes to mitigate the consequences of past errors; and providing full information about preventive measures.

The publication of more detailed and more accurate information should not be viewed as a sufficient condition for overcoming public opposition to a hazardous waste facility. On the other hand, it is a necessary ingredient in a responsible siting process. The development of effective ways of communicating uncertain risks, especially when they involve low probability events or the effects of chronic exposures, is obviously difficult. The distrust of institutions responsible for providing the information and the emotions surrounding involuntary technological risks add significantly to the problem. Difficulties in risk communication about radioactive wastes have been evident in most recent siting efforts. The use of abstract technical terms, the use of quantitative risk information, and the inadequate use of metaphors to explain risks have all contributed to poor communication (see, e.g., van der Pligt, 1988, 1989; see also Chapters 2 and 8).

Another contributing factor is the inadequacy of existing procedures for dealing with the public. Too often authorities rely upon public hearings at which 'experts' with little experience of risk communication are confronted by angry, upset local citizens. Public hearings are a notoriously poor vehicle for either the transfer of information or the involvement of interested publics (Checkoway, 1981a,b; Vlek, 1986). Statements and assessments of environmental impact tend to be extremely complicated and fail to convey the relevant information to the lay public. Too often reports from prospective site-developers or relevant agencies are characterized by self-interest or justification of actions taken (cf. Kasperson, 1985). It needs to be added, however, that the current understanding of how best to convey risk information is limited (see Chapter 8). Officials of regulatory agencies find it difficult to present risk information to lay people. Most people find it difficult to interpret probabilities or compare 'familiar' risks (for example being hit by lightning) with those of recently introduced technologies (for example increased cancer incidence due to a nuclear accident). Slovic, Fischhoff and Lichtenstein (1980) have pointed out that presenting a fault tree (a tree-like structure summarizing the possible causes of an accident and their probabilities), designed to demonstrate the small probability of the risks, may instead impress the observer with the large number of potential things that can go wrong.

Public participation Successful siting requires a conception of the public as a partner rather than as an adversary. Public involvement in monitoring and control has been discussed in the previous sections. Involvement in the decision-making process also seems to be a necessary ingredient of a more successful siting policy. Unfortunately this is not a simple matter. Limited experience so far with increased public involvement in policy decision-making has provided no clear answers. On the one hand, findings reported in the previous sections suggest that characteristics of the decision process are of extreme importance. Increased participation could improve local acceptance and help to reduce existing distrust. Some experiments, however, have failed to produce this. Both the public inquiry system in the UK and the 'National Nuclear Debate' in the Netherlands have not led to increased public acceptance. Both were/are expensive exercises with mixed success (see, e.g., Wynne, 1982; Vlek, 1986). Too often, public participation is used as a last resort when there is a political stalemate. Investigation of the trust and credibility attributed to these and other attempts to involve the public seems necessary before there is large-scale implementation of these programmes. The basic problem seems to be that public discussions become extremely technical and hence incomprehensible to lay citizens.

Conclusions

This chapter has provided an overview of the rather unfortunate history of siting nuclear waste facilities. Finding storage for the increasing quantities of low, intermediate, and high-level wastes has become an urgent matter, given that most temporary on-site storage facilities will soon be full. Finding means for the disposal of radioactive wastes that are both technically and socially acceptable is one of the major challenges facing the nuclear industry. The expansion of the number of nuclear power stations has come to a standstill in most Western countries but this has not resolved the radioactive waste problem. Since around 1960 considerable amounts of waste have been generated and there is still no agreement on possible solutions that are both technically and socially acceptable.

It is also clear that finding acceptable solutions will be complicated and that these need to be based on multidisciplinary approaches to the issue. Social psychology and sociology have contributed by stressing the importance of risk perception and communication and pointing at the crucial role of equity. Translating these insights into adequate policy solutions will only be possible with the help of political and management sciences. Improved communication and increased public involvement and participation seem key ingredients of this process.

6

Environmental stressors

This chapter discusses the literature about how people deal with stressors; i.e., circumstances that have negative effects on their performance and well-being. The processes by which people react to threatening environmental events will be considered as a form of 'stress'. Several serious nuclear accidents (e.g., Three Mile Island and Chernobyl) are commonly accepted to have been stressful. Other threatening and stressful situations produced by nuclear energy are the fear of health and economic problems caused by routine radioactive emissions and, most often, the disruption caused by siting procedures for nuclear facilities. Organisms under stress must resist, adapt or otherwise find ways of coping. If coping is impossible and the stress is intensive and persistent, negative consequences will follow, such as physiological or psychological discomfort and, in some cases, ill health.

Traditionally, theories of stress have followed one of two approaches, using either a physiological or a psychological perspective to conceptualize stress-related processes. In this chapter I will give a brief overview of both approaches. After a definition of the concept of stress, I will describe the characteristics of environmental stressors. This results in a classification of four types of stressors, two of which are most relevant to issues discussed in this book, i.e., cataclysmic events and ambient stressors. I will then briefly describe the two theoretical perspectives on stress: the physiological approach and the psychological approach. This general background information is followed by a brief review of the effects of stressors. This overview will then be applied to the issue of nuclear energy. In this way I will attempt to demonstrate the usefulness of the stress concept to help improve our understanding of the behaviour of both lay people and experts when dealing with issues related to nuclear energy.

Definition of stress

Early definitions of stress varied in the degree to which they emphasized the responses of the individual or the context that caused disruptions of behaviour. Appley and Trumbull (1967) and Evans and Cohen (1987) have summarized several objections to these approaches. The first objection to response-based definitions of stress concerns the relative lack of attention to important temporal parameters in stress. A focus on outcomes tends to ignore the fact that highly variable situations can lead to similar outcomes. Furthermore, other sources of stress, cultural or social norms and social support may all influence responses to specific stressors (e.g., Kaplan, 1983; Pearlin, 1982). Finally, there is a considerable lack of correspondence among response measures. It has proved rather difficult to select responses that invariably occur when adaptive resources are tested.

Situation-based definitions of stress, on the other hand, have been criticized because of significant variations in individual responses to the same situation. Past history, the appraisal of a threat, and coping styles all vary between people. Furthermore (as noted by Evans and Cohen, 1987), with the exception of very extreme stimuli, no stimulus is a stressor to all persons. Similarly, the same stimulus does not always lead to the same response from a particular individual. Not surprisingly, it has also proved rather difficult to define situations in terms of relative 'stressfulness'. Moreover, the emphasis on the situational context makes it very difficult to conceptualize the immediate and long-term consequences (behavioural, psychological and physiological) of stress (see also Evans and Cohen, 1987).

The above problems have led most stress researchers to adopt a relational, interactive definition of stress. According to this perspective, stress is a process that occurs when there is a perceived imbalance between situational demands and response capabilities of the organism (see, e.g., Lazarus, 1966). An important aspect of this relational perspective on stress emphasized by Lazarus and his colleagues is that for stress to occur, the individual must perceive this imbalance. In other words, stress occurs when one decides that situational demands are likely to test or even exceed one's personal coping capacities.

Characteristics of environmental stressors

In general the literature distinguishes four types of environmental stressors: cataclysmic events; stressful life events; daily hassles; and ambient stressors (see, e.g., Baum, Singer and Baum, 1982; Campbell, 1983; Evans and Cohen, 1987; Lazarus and Cohen, 1977).

Cataclysmic events are sudden catastrophes that demand major adaptive responses from those directly affected. In this chapter I will review several examples of nuclear accidents that fall in this category. Other examples of cataclysmic events include earthquakes, volcanic eruptions, hurricanes and discoveries of contaminated soil in residential areas.

Stressful life events are major incidents in the lives of people requiring major adaptive responses. This category includes such things as major changes in relational conditions (marriage, divorce, death), or major changes in economic conditions (for example the loss of a job). This type of stressor will not be discussed in the present chapter.

Daily hassles are the typical events of ordinary life that may cause tension or irritation. Examples include environmental events, such as rush hour traffic, noisy neighbours, and long queues in the supermarket; work issues such as disagreement with colleagues and excessive workload; or interpersonal problems such as disagreement with a spouse. Daily hassles tend to be more common and short-lived than life events. This type of environmental stressor is also less relevant in the present context.

The term ambient stressors has been introduced to distinguish more continuous, stable and relatively intractable conditions of the physical environment (cf. Campbell, 1983). Most ambient stressors are background conditions. Examples include traffic noise, air pollution, poor climate control and bad lighting. Individuals living with chronic air pollution or traffic noise, for example, are likely to habituate to these environmental conditions. More active coping responses in these conditions are often not feasible or are expensive costs (e.g., moving house). Such responses are less attractive than accommodating to suboptimal living conditions. Living near a nuclear power station or a nuclear waste facility has also been linked to this category of ambient stressors (see, e.g., van der Pligt, 1985; van der Pligt, Eiser and Spears, 1986a). Another example is the long uncertainty imposed on communities when they are selected as a possible site for a nuclear facility.

In this chapter I will focus on stressors falling in the first and last categories; i.e., cataclysmic events such as nuclear power accidents and ambient stressors such as living near a nuclear power station during the construction phase or living near an operational nuclear facility.

Several characteristics of stressors influence individual reactions to stressful events (see Evans and Cohen, 1987). A first characteristic is the degree to which a stressor is perceptually salient or easily identifiable

(Baum, Singer and Baum, 1982; Campbell, 1983). A related factor concerns the duration and periodicity of environmental stressors. Duration refers to both the extent of a person's previous exposure to the stressor *and* the length of current exposure. Periodicity refers to the predictability or regularity of the stressor as well as its continuity. Some stressors are discrete (such as a major accident, the discovery of nuclear contamination in one's neighbourhood); others are continuous (such as worries about routine-level radioactive emissions, traffic noise in the construction phase of a nuclear power plant). Adaptation processes will be strongly affected by both duration and periodicity.

A second group of characteristics is related to the type of adjustment required by the environmental stressor. Environmental conditions that are very intense and uncontrollable are likely to lead to accommodation and emotion-focused coping rather than efforts to deal with the stressor directly (see, e.g., Lazarus and Cohen, 1977). These coping and adaptation processes may in turn influence the psychological, psychosomatic and health consequences of exposure to that stressor.

In describing stress-related reactions controllability is an important factor. Control can be described as a psychological (appraisal) process that is determined not only by the stressor but also by individual characteristics and personal resources for coping. Control can also refer to the possibility of influencing a specific situation; i.e., the occurrence, duration and periodicity of an environmental stressor. In this sense control refers to characteristics of the situation. As was discussed in previous chapters, controllability plays an important role in risk acceptance. Given the issues described in Chapter 2 it is not surprising that uncontrollable stressors are typically seen as more threatening, at least initially. Furthermore, uncontrollable stressors are frequently associated with negative effects on well-being, health and behaviour (Baum, Singer and Baum, 1982; Cohen, 1980). If a stressor remains uncontrollable for a considerable length of time, it is more likely to become an unnoticed, background characteristic as a result of habituation processes (Campbell, 1983). Habituation is more likely to be the response to moderate than to extreme stressors. Accommodation to stressors which are difficult to change through instrumental coping efforts has been noted in research on coping with various sources of pollution (such as noise and air pollution) and technological risks (e.g., Baum, Singer and Baum, 1982). When an aversive situation cannot be modified or eliminated, the most feasible option is often some form of denial or reappraisal of the stressor (Folkman and Lazarus, 1980; Pearlin and Schooler, 1978).

Related to controllability is the predictability of stressors. Some environmental stressors are more predictable than others which will influence both the manner in which they are coped with and the way they influence health and well-being. For example, habituation to continuous noise (e.g., motorway traffic) or predictable noise (e.g., railway traffic) is common and can be more easily achieved than habituation to, for instance, building noise, which is intermittent and less predictable.

Finally, the acceptability of environmental stressors is also influenced by factors such as the perceived necessity and importance of the activity or event. Situations that are seen as necessary and/or important tend to be accepted more readily and hence cause different kinds of reactions. For instance, aircraft noise caused by a sudden increase of flights with emergency goods for a disaster area is likely to lead to different reactions than the opening of a new disco in a quiet neighbourhood.

Theoretical perspectives on stress

Reviews of research in environmental stress usually classify work as falling within one of two research traditions: the physiological tradition and the psychological tradition. As was argued by Evans and Cohen (1987), these theoretical paradigms are not necessarily contradictory, but rather emphasize different dimensions of stress-related processes. In the brief overview of theoretical approaches to environmental stress, presented in this section, I will follow Evans and Cohen and describe these traditional approaches first. I will then discuss more specific issues concerning the environment and the individual that lead to a stress response.

The physiological approach   Selye's (1956) work on stress is perhaps the best-known physiological perspective. His model emphasizes the physiological responses of an organism to noxious stimuli. The model concentrates on homeostatic processes, by which the body responds to aversive conditions that disrupt the internal equilibrium. Selye found that animals responded to a variety of aversive stimuli (e.g., heat, noise, irritants) with similar changes. These changes consisted of three responses: enlargement of the adrenal glands, shrinkage of the thymus, and ulceration of the gastrointestinal tract. According to Selye these responses were non-specific; i.e., all noxious stimuli produced the same set of responses. Selye further proposed a three-stage general adaptation syndrome (or GAS): alarm, resistance and exhaustion. The first stage (alarm) is characterized by recognition of the threat and preparation to resist it. After the organism has become aware of the stressor, it prepares a response by increasing vital functions such as adrenal activity, respiration and cardiovascular activity. When the various systems have been mobilized the organism enters the state of resistance. This will continue until reserves have been depleted or the stressor is overcome. If the stressor persists or coping abilities are low, the organism enters the stage of exhaustion. This is associated with diseases of adaptation such as immune deficiency and cardiovascular diseases.

Evans and Cohen (1987) emphasize three specific implications of the physiological stress model. First, if various stressors cause identical, non-specific responses characterized by the tripartite response syndrome, stress is likely to be additive. Responses to a specific stressor will thus be influenced by both the severity of the specific stimulus or situation as well as by other recent or contemporaneous stressors (see also Fleming, Baum and Singer, 1984). Secondly, some adverse health effects can result from the adaptation processes themselves. The energy needed to adapt can affect the cardiovascular system and increase susceptibility to infectious diseases. Thirdly, the human body has a finite amount of adaptive energy. When this has been exceeded, adverse consequences are likely to occur (see, e.g., Glass and Singer, 1972).

One of the less developed aspects of Selye's model of stress are the mechanisms that trigger the general adaptation syndrome. Empirical evidence suggests that the syndrome is triggered only when the person *perceives* threat or psychological harm (Mason, 1975). In other words, cognitive appraisal of the threat seems necessary to trigger a stress response. Whether this is indeed the case is a point of controversy in the stress literature (Mason, 1975; Selye, 1975). As was noted by Evans and Cohen (1987), there is also increasing evidence that both cognitive appraisal and coping processes can influence physiological responses to stress. Different coping efforts cause different physiological profiles. All of these findings point to the necessity of including psychological factors in the study of stress-related processes.

The psychological approach This approach deals primarily with the individual's *interpretation* of specific environmental events and combines this with an appraisal of personal coping resources (Lazarus, 1966). Primary appraisal refers to the process of evaluation of the stressor and depends on the situation and the person. Situational variables include the imminence and likely magnitude of harm; the ambiguity of the stressor; the duration, periodicity and predictability of the stressor; and, finally, the extent to which the stressor can be controlled. Personal factors that influence primary appraisal include beliefs about self-efficacy (the idea that one can exert control over one's environment or the stressor) and the importance of the goals or needs threatened by the stressor. After the initial evaluation of a situation in terms of threat, harm, or challenge, secondary appraisal processes come into play. During this phase the individual selects and evaluates coping resources in order to deal with the stressor. One general classification of coping responses is problem-focused coping versus emotion-focused coping. The first class of coping strategies involves attempts to change the situation to reduce aversive impact. Emotion-focused coping, on the other hand, is less instrumental and refers to changes in the individual's internal responses to the stress-inducing situation.

Thus, according to the psychological perspective, stress occurs when a situation is *perceived* as demanding and as exceeding one's coping resources. The individual's perception of the imbalance between environmental demands and personal coping resources is the critical variable in determining the nature of the stress response. This implies that stressful situations are not uniformly aversive. Personal and social mediators can reduce or enhance the effects of potentially stressful situations. This mediation can occur by influencing either one or both of the appraisal processes (cf. Evans and Cohen, 1987). For example, perceived control over a stressful situation may make it seem less threatening (primary appraisal) and/or affect the individual's perception of the number of options available to cope with the stressful situation (secondary appraisal). According to the psychological perspective, stressors can affect the individual in a wide variety of ways. These effects include self-reports of stress and related symptoms (e.g., nervousness, tension, anxiety), negative affect and interpersonal behaviours, deficits in information processing and reduced task performance.

One of the more specific issues in this field of research concerns adaptation and coping. Glass and Singer (1972) and Cohen (1980) have put forward the view that cognitive fatigue may be a cumulative cost of adapting to stressful situations. Negative after-effects of prolonged coping efforts also include reduced tolerance for adverse circumstances and suboptimal cognitive performance. Examples of the cumulative effects of coping with environmental stressors have been found after exposure to noise. As was argued before, the resulting fatigue may also reduce the capacity to cope with subsequent stressors.

Coping responses themselves also have direct physiological effects. When effort is needed to maintain task performance during stress, cardiovascular activity greatly increases. In the absence of relevant feedback to the organism about the efficacy of his or her coping attempts, even more physiological activity results. Chronic adaptive efforts may lead to disease, either directly as in the case of increased cardiovascular activity, or indirectly, as a result of reduced immunological defences to infectious diseases. As Evans and Cohen (1987) argued, more research is needed on both the precise mechanisms of stress–disease links and the physical, social and psychological characteristics of situations that enhance or reduce coping activities that are successful.

Research in the 1970s and 1980s has tended to focus on situational and individual factors that influence the choice of coping strategy. Most of this work attempts to understand the occurrence of maladaptive and adaptive coping responses. In this context it is important to distinguish between problem-focused and emotion-focused coping. This distinction is related to Leventhal's (1970) fear-drive model in which he distinguishes fear-control from danger-control. As we will see later, Janis and Mann's (1977) and Rippetoe and Rogers' (1987) work on coping provides further insights into the effects of coping strategies on human behaviour. Maladaptive coping styles such as defensive avoidance, fatalism, hypervigilance and wishful thinking are most likely to occur when one's control over the stressor is minimal. To this issue of control we turn next.

A concept of particular relevance to the present discussion is that of psychological control. There is ample evidence that human beings have a strong need for control (Averill, 1973; Langer, 1975). Adverse consequences associated with lack of control include negative feelings and emotions, reduced performance and errors and reduced efficiency in cognitive tasks. Actual or perceived control over a stressor generally leads to fewer negative consequences than exposures to stressors that are uncontrollable. A substantial amount of research shows that uncontrollable or unpredictable environmental stressors cause greater stress in human beings. Work on noise (Cohen and Weinstein, 1982), air pollution (Evans and Jacobs, 1982) and toxic exposure (Renn, 1982) have found significant reductions of negative consequences if control over the stressor was available.

Chronic exposure to uncontrollable environmental stressors may also produce greater susceptibility to learned helplessness (cf. Seligman, 1975). If one cannot predict or assert control over a particular stressor, one may generalize from this and assume that behaviour can have little influence on environmental outcomes. Generalizing from an assessment of low ability can have important behavioural and psychological consequences, as was shown by Seligman. Reduced well-being and even long-term negative health consequences may be related to learned helplessness.

Another effect of prolonged exposure to aversive, uncontrollable stressors may be a change of emphasis from problem-focused coping strategies to emotion-focused coping. One way to control a threat would be to reappraise it. This happens when an aversive condition (e.g., living near a toxic waste facility) that was initially viewed is a critical way is redefined as being a minor problem. Denial of harmful effects may also occur as a result of experience with an uncontrollable, ambient stressor (Campbell, 1983). All in all, there is considerable evidence that chronic exposure to environmental stressors can cause negative reactions because of restrictions in perceived control. Furthermore, provision of real or perceived control over stressors is likely to reduce the adverse consequences of environmental stressors for human health and behaviour.

The concept of predictability is related to control and tends to have similar effects. Situations that are unknown, highly ambiguous or difficult to interpret may be stressful. For instance, when one fails to discern the meaning or function of an object or a situation, confusion and uncertainty as well as stress may occur. Unpredictable stressors may also be more aversive because individuals are left without signs indicating the absence or presence of the aversive situation; i.e., it is impossible to estimate when the situation is safe (Seligman, 1975). When confronted with a predictable stimulus condition, one can at least relax occasionally and thus achieve some recovery during safe periods.

The effects of stressors

In this section I will briefly summarize the range of stressor-effects noted in the previous sections. This overview of stressor-impacts will then be related to public reactions to nuclear energy. In this way I will attempt to investigate the usefulness of the stress paradigm for understanding how various aspects of nuclear energy affect human health and well-being. Following Evans and Cohen (1987) I will focus on four specific areas of stressor impacts.

Physiological effects Following physiological models of stress, stress in human beings has been measured by examination of various endocrinological responses. There is substantial evidence that a wide variety of aversive stimuli cause increased catecholamine and corticosteroid output that is detectable either in blood or urine (see, e.g., Baum, Grundberg and Singer, 1982). These hormones, adrenalin in particular, produce changes in various organs related to increased sympathetic arousal. To illustrate the latter, numerous investigations have recorded psychophysiological indices of stress, including increased blood pressure, heart rate and respiration rates. Other indices are muscle tension, and skin conductance (see Baum, Grundberg and Singer, 1982).

Task performance The influence of stressors on human task performance remains extremely difficult to characterize. The primary reason for this is that most people can, at least temporarily, overcome the aversive effects of a stressor by coping devices, such as increased effort or concentration. Nevertheless, there are certain patterns of reduced task performance that occur under stress. These are most obvious when the task requires rapid detection, sustained attention, or attention to multiple sources (see Evans and Cohen, 1987).

Janis and Mann (1977) focus on conditions that determine whether stress will facilitate or interfere with adequate information processing. Their work applies mainly to (emergency) decision-making in the face of threatening situations, oncoming disaster and/or fear-arousing messages that urge protective action to avert serious threats or health hazards. They distinguish between adaptive and maladaptive coping styles. The latter (e.g., defensive avoidance, hypervigilance) can be characterized by inadequate search for information, uncarefully thinking through the consequences of alternative courses of action, distortion of the meaning of information and selective forgetting.

Research suggests that stress causes premature closure; i.e., decisions are made before all relevant data have been considered. Related shortcomings in decision-making under stress include fixation on a limited number of aspects of a specific problem or task. Similarly, novel information or tasks requiring novel approaches are more likely to be redefined in terms of pre-existing theories, habits or schemata. Implications of these effects of stress will be discussed later in this chapter.

One of the negative consequences of stress on information-processing mentioned by Janis and Mann (selective forgetting) is supported by research on memory under stress. Under increased levels of stress several memory deficits have been found: memory for incidental or secondary information in a task tends to be poorer. Stressors also cause faster processing of information in working memory but this increased speed is obtained at the expense of total capacity. Finally, there is also evidence of poorer comprehension of complex information. This reduced performance is also associated with reduced working memory capacity (Broadbent, 1971).

Affect and social behaviour Both self-reports of affect and interpersonal behaviours such as aggression indicate effects of stressors. Many studies have demonstrated greater anxiety, tension and nervousness under aversive stressful conditions (Lazarus, 1966; McGrath, 1970). An important methodological issue with respect to self-reports of stress is whether individuals can validly report the degree of stress they are experiencing. This issue needs further research, but according to the psychological perspective on stress, perceived stress rather than some objective indicator is the driving force behind responses to an aversive situation. This perspective thus accepts self-reports as valid indicators of stress. One possible exception concerns strategic responses: if reporting stress-related complaints could lead to policy measures to reduce the impact of a stressor one would expect a tendency to overestimate the frequency and intensity of stress-related complaints.

Some research has found more negative social behaviour under stress. Examples include less altruism and cooperation and increased competitiveness, hostility and aggression. Aggressive behaviours are typically studied in laboratory settings (games, mock learning experiments) where aggression is mobilized as the willingness to deliver punishment (e.g., electric shocks) to fellow subjects. Overall, aggression tends to be more pronounced when subjects have been previously angered or exposed to aggressive, modelling behaviours (Cohen and Spacapan, 1984). Helpful behaviour has also been directly measured by observing reactions to a person in distress or by assessing the willingness to cooperate with requests for aid. In general, cooperation and helping behaviour are found to decrease under stress (Cohen, 1980; Evans, 1982). This reduced willingness has been found in both laboratory and field research (e.g., as an effect of high noise levels in urban environments).

Adaptation, well-being and health If people can adapt to stressors through coping mechanisms, some cumulative costs may become apparent after prolonged stressor exposure. Adaptive behaviours may often reduce the immediate stress response in the form of habituation, but the process itself may in turn take its toll. These negative aftereffects of coping may include less ability and motivation to cope with subsequent stressors; social and emotional adjustment problems; and even greater susceptibility to infectious diseases (Cohen, 1980; Evans and Cohen, 1987; Glass and Singer, 1972). Another important type of residual coping behaviour is the tendency to overgeneralize coping responses. An example of overgeneralization mentioned by Evans and Cohen (1987) is learning to cope with loud noise by tuning or filtering out auditory stimulation. Evidence indicates that tuning out becomes a routine part of the cognitive repertoire of persons chronically exposed to noise even when they are in quiet conditions (cf. Cohen et al., 1980).

Similarly, people could feel helpless even in situations where they can influence the outcome.

A final group of after-effects from chronic exposure to stressors includes effects on health and psychological disorders. When adaptive resources are strained over long periods, some adverse health effects are likely. Three types of physiological effects associated with coping with chronic stressors are cardiovascular disorders, gastrointestinal problems and lowered immunological resistance to infectious diseases. As was noted by Evans and Cohen (1987), some progress has been made in identifying the mechanisms of these effects, but knowledge of the exact nature of these processes is still limited. Psychological effects include reduced well-being (anxiety, tension, nervousness) and also the aforementioned motivational consequences caused by a feeling of helplessness.

The literature on stress suggests a wide variety of immediate and long-term effects of stressors. Although the picture of possible after-effects is becoming clearer, knowledge is still limited. There are several obvious reasons for this. It is difficult to investigate systematically the effects of acute and intense stressors in the laboratory. (Ethical problems are also a reason for the absence of a vast literature on these stressors.) Moderate stressors, on the other hand, can easily be investigated in the laboratory but studies of these tend to focus on immediate and short-term effects. It can be argued that the more severe consequences of prolonged exposure to moderate stressors will be long-term.

The consequence is that both acute major stressors such as cataclysmic or life events and the long-term effects of moderate stressors such as aircraft noise have mainly been studied in the field. Unfortunately, while we know a considerable amount about immediate effects, we know hardly anything about medium- and long-term effects of environmental stressors. As was argued before, more research is needed on the nature of stress–disease links and on the physical, social and psychological characteristics of stressful situations that enhance or reduce successful coping behaviours. Notwithstanding the gaps in our knowledge, the concept of stress seems important in the context of nuclear energy. To this issue I will turn next.

Nuclear energy and stress

This final section discusses the relevance of the stress concept to nuclear energy. It attempts to show that nuclear facilities could have characteristics that are likely to enhance stress in communities close to a nuclear facility. Too often the characteristics of policy decision-making processes tend to enhance stress. These situations can be summarized under the heading ambient stressors. Research in the context of nuclear energy tends to focus on cataclysmic stressors, i.e., serious nuclear accidents. I will consider these first.

Nuclear accidents Several accidents, such as the Three Mile Island release of radioactive materials in 1979, have had an enormous impact on the public. The impact has included economic, social and psychological effects. Most of the immediate stress-related reactions were similar to those described earlier under the heading cataclysmic events. As we will see in Chapter 7, a considerable amount of research has been conducted around Three Mile Island. It found significantly high levels of stress in local communities near to the reactor site over two years after the accident. Furthermore, coping styles were also related to stress responses. Subjects who opted for problem-focused coping styles reported more stress-related complaints than those who opted for emotion-focused coping styles. In this research, conducted by Baum and his associates, a variety of indices were used (self-reports, cognitive functioning and physiological responses) (e.g., Baum, Gatchel and Schaeffer, 1983).

The research found significant increases in stress-related symptoms that were likely to last for quite a while. The main reasons appear to have been the prolonged economic uncertainty (e.g., lower house prices, resulting in lower mobility) and uncertainty about delayed health effects. It is likely that similar types of effect were even more severe and widespread after the Chernobyl accident in 1986. Not much is known about the affected areas in Ukraine and some parts of Europe, but newspaper reports provide anecdotal evidence from several countries. Acute stress resulted in panic and extreme behavioural responses, including higher abortion and suicide rates. These reactions, however, were the exception rather than the rule (see Chapter 7).

The detection of contamination as a result of routine emissions from nuclear power stations is bound to have similar, if less extreme, effects as those described above. For instance, media reports of emissions at Sellafield in Northwest England provided numerous indications of increased stress (see also van der Pligt, Eiser and Spears, 1987b). This case had similarities with the major accidents. People feared extreme economic consequences (such as a reduction of house prices) and uncertainty about health impacts. They were particularly concerned about the reported increased rate of leukaemia cases.

Accidents like those at Sellafield can be compared with cases involving toxic waste, such as the leaking chemical waste dump at Love Canal in western New York (Levine, 1982), or the detection of contaminated soil in several communities in the Netherlands (see van der Pligt and de Boer, 1990). The immediate consequences of such cases are quite severe and are seen as a threat to one's existence. Home-owners fear substantial financial losses and reduced mobility since they are unable to sell their home for an acceptable price (see van der Pligt and de Boer, 1990). For people who stay in an area, possible long-term health consequences remain a source of worry.

The immediate impact of reduced house prices could be alleviated by guaranteeing fixed house prices in the case of an accident. But accidents can have more generalized economic consequences for areas affected. These include reduced willingness of companies to settle in these areas and possibly reduced willingness of people to migrate to them. These generalized effects are unlikely to increase the stressfulness of the situation, but they are also unlikely to improve the general situation in the area. In the case of possible health problems, adaptive and maladaptive coping styles and their consequences for psychological well-being and health are important areas of research that need more attention. Indications of research in other areas suggest that groups already at risk are most vulnerable to long-term consequences. Programmes aimed at changing maladaptive coping styles could play an important role in reducing these adverse health effects.

Nuclear accidents will be discussed in more detail in Chapter 7. The question that concerns us here is the usefulness of the stress concept in the study of human reactions to nuclear accidents and routine emissions. This seems to be threefold. First, major accidents do have a clear impact on perceived stress and stress-related complaints. These adverse consequences for public health should be taken into account in risk assessment and impact assessment. The possible long-term consequences of having to cope with these adverse circumstances (i.e., prolonged uncertainty) need more attention. The immediate impact of reduced house prices could be alleviated by guaranteeing fixed house prices in case of an accident. Accidents can have more generalized economic consequences for areas affected by a nuclear accident such as reduced willingness of companies to settle in these areas. These consequences also have an adverse impact on the general situation in the area.

Secondly, nuclear accidents show several characteristics that make them extremely stressful. The visibility and salience of the stressor are quite high for those living nearby. Controllability over the consequences is generally perceived to be extremely low and the likely severity of possible consequences is high. After an accident the duration of the stressful situation is potentially unlimited. Sometimes these unfortunate stimulus characteristics are enhanced by ambiguity in the information provided to those involved. Reduction in stress-related complaints could be obtained by the fast provision of clear information. Stressful situations and perceived harm or threat are likely to polarize public debates and this seems exactly what has happened too often in the context of risk management after major accidents. Thus, some of the important antecedents of stress reactions seem an integral part of this new technology. It is unfortunate, however, that these effects are further enhanced by inadequate information programmes (see also Chapter 8).

Thirdly, the application of the stress paradigm to nuclear energy issues underlines the importance of undertaking further research into the effects of cataclysmic events. Early findings in this research area suggest both medium- and long-term consequences for well-being and health. Both experimental and longitudinal epidemiological research are badly needed to improve insight into stress-related processes, the role of coping and consequences for health and well-being.

Stress research can also be applied to a different aspect of nuclear accidents, i.e., emergency behaviour. Evidence suggests that stressful conditions produce decreased quality of information processing and performance in general. The implications are straightforward and simple. Firstly, workers should be trained to cope under such circumstances. (Too many accidents are the result of human error, which are compounded by more errors made under stressful circumstances.) The design of complicated facilities such as nuclear power stations should be such that human error is prevented or at least unlikely. For instance, it has been argued that poor ergonomics contributed to the Three Mile Island accident, by making difficult the initial responses to the accident. The decrease in the quality of information-processing that results from stressful conditions also has clear implications for the provision of information to the public in case of an emergency. Clear, simple guidelines, coming from one source, distributed via several channels (brochures, mobile loudspeaker units, local radio and television) are essential to improve the effectiveness of emergency behaviour programmes (see also Chapter 8).

Daily hassles and ambient stressors are concepts that apply to other aspects of nuclear energy. Daily hassles are experienced by communities confronted by the construction of a nuclear power station. Local life is disrupted, by both increased traffic noise and social changes (e.g., a sudden influx of workers into the area). The concept of ambient stressors is most relevant to communities shortlisted as a possible future site for a nuclear facility and communities in the vicinity of an operational nuclear power station or waste disposal site. The severity of the resulting stress-related complaints is bound to be less extreme than in the case of accidents, but should not be underestimated.

Stress caused by a prolonged threat, such as being on a shortlist for the siting of a nuclear power station or hearing rumours about possible effects of routine emissions, has not been systematically investigated. It is likely to be similar to the stress found where contaminated soil had been detected in residential areas. In these circumstances about 10–15 per cent of residents reported stress-related complaints. Most of these, however, were not related to worries about possible adverse health effects or economic effects but to the ambiguity and inconsistency of the authorities' reactions. The totally different time-scales of regulatory agencies and other authorities tend to prolong a situation of considerable uncertainty for the citizens involved and are often interpreted as lack of urgency. This, in turn, tended to induce chronic stress and enhance stress-related complaints. It also led to increased involvement of local citizens and to emotional confrontations between the involved parties. Again, inadequate communication and different time-scales not only enhance stress-related complaints but also tend to polarize further the nuclear debate and hinder efficient policy decision-making.

The area around Sellafield in the UK is an example of a chronic stress situation. For several years inhabitants have been confronted with an extremely uncertain situation about the possible adverse health effects caused by routine emissions from the nearby nuclear facilities. Unfortunately, to my knowledge, there has been no systematic research on the effects of this prolonged stressful situation.

What can be done to alleviate the effects of daily hassles and ambient stressors? Evidence suggests that daily hassles are likely to lead to stress-related complaints and that these are accompanied by reduced acceptance of the construction of nuclear power facilities. Increased awareness of these understandable reactions could encourage special provisions, such as the building special routes into reactor sites to minimize disruption from traffic.

The effects of ambient stressors are more difficult to investigate but research in different domains (noise, air pollution) indicates short-term effects. Knowledge about long-term effects is limited. The stressfulness of such situations is, however, partly a function of the way authorities and regulatory agencies operate. Too often they follow a long time-scale, which is unfortunate for citizens involved. Hence, unwittingly, they increase stress and decrease the acceptance of the construction and operation of nuclear facilities. Prolonged periods of uncertainty are bound to enhance stress-related complaints; too often the length of time needed to reach siting decisions and the ambiguous information provided to the public are likely to increase stress and decrease public support.

A final remark about these moderate stressors concerns our lack of knowledge about long-term effects. Immediate effects are well-documented but psychological and health-related effects of continuous exposure to these stressors need more research.

Stress and policy decision-making Major findings of the stress literature can also be applied to nuclear decision-making. First, the adverse effects of uncertainty and ambiguity over prolonged periods of time should be taken into account. There should be a faster provision of less ambiguous and/or contradictory provision. This applies to both the after math of nuclear accidents and siting procedures. Secondly, the seriousness of possible long-term effects of severe stress should be taken into account in cost–benefit analyses of nuclear power. Finally, it seems that nuclear decision-making at all levels is frequently taking place under increased levels of stress. It seems wise to try and minimize the effects of stress on decision-making. For emergency situations this means better training, improved design and better information for both workers and the relevant risk-management authorities. At a more general level, this also means that policy decision-makers should be aware of the possible impact of stress (caused by public opposition) on the quality of their own decision-making. The possibility of biased information-processing, biased memory and distorted interpretations of facts does not only apply to lay people or affected citizens but also to policy decision-makers.

Conclusions

This chapter has described a series of stress-related reactions that are relevant in the context of nuclear energy. It seems that the nature of these reactions and their immediate and long-term consequences are (a) not fully understood and (b) not always taken into account in nuclear policy decision-making.

Some measures (consistency and speed of information provision) could help to reduce stress-related complaints in both siting procedures and after nuclear accidents. Similarly, simple measures could reduce the impact of stressors during the construction of nuclear facilities. Denial of the stressful characteristics of nuclear technology during siting procedures and after accidents has resulted in a situation in which the relevant authorities have often added to the stressfulness of specific situations. This is one of the causes of the increased polarization in the nuclear debate. Much effort will be needed to restore some balance. This polarization is most evident in situations where there is a possible serious health threat. The long-term consequences of these more extreme stressors need more research.

The stress concept can also be applied to the design and lay-out of nuclear facilities; improved ergonomics could prevent future accidents. To this and related issues I will turn in the next chapter, in which the nuclear accidents at Three Mile Island and Chernobyl will be discussed.

7

Nuclear accidents: Three Mile Island and Chernobyl

Technological disasters are less familiar to most people than are natural disasters, partly because they happen less often. Like natural disasters, technological disasters include events that are powerful and sudden (e.g., industrial accidents, bridge collapses and air crashes). They also include events in which highly toxic substances are released into the atmosphere. This chapter discusses the impacts of nuclear accidents. Most attention will be paid to large-scale nuclear accidents involving the release of radiation; i.e., the accidents at Three Mile Island and Chernobyl. A number of themes discussed earlier in this book, such as public reactions, environmental stress, attitudes and risk acceptance, will again be considered but now in the context of nuclear accidents.

Although technological catastrophes have only recently become a focus of attention, there are a few studies that investigate their effects. As was noted by Baum, Fleming and Davidson (1983), most research, however, does not compare accident victims with a control sample of unaffected people. Moreover, it tends to rely on self-report measures and is sometimes based on inadequate sample sizes. Notwithstanding these shortcomings, the available data form a consistent pattern suggesting the recurrent effects of exposure to technological disasters.

Several studies of technological catastrophes have examined their short-term effects. Adler (1943), for example, interviewed survivors of a large fire, in which almost 500 people died. He found that over the eleven months after the fire, nearly one-third of the survivors displayed nervousness or anxiety. Adler also reported that 60 per cent of those survivors who did not develop psychiatric complications had lost consciousness during the fire for a prolonged period. Only 12 per cent of those who did develop psychiatric problems had had prolonged loss of consciousness. This suggests that experiences during the fire were primarily responsible for the subsequent difficulties.

Other studies have focused on the longer-term effects of disasters (for a brief overview see Baum, Fleming and Davidson, 1983). Leopold and Dillon (1963) reported the results of a longitudinal study of survivors of a marine explosion. They found that three-quarters of the survivors had received some help for psychiatric complaints and that nearly all of them had experienced work-related problems. Several studies of the dam collapse and flood at Buffalo Creek, West Virginia, USA, in 1976 revealed a substantial amount of disruption among survivors. Titchener and Kapp (1976) reported high rates of emotional disturbances – including anxiety and depression – more than two years after the incident. Gleser, Green and Winget (1978; 1981) investigated flood survivors and found evidence of continued anxiety, depression and hostility two years after the incident. This was not the case with subjects from a control group drawn from non-flooded parts of the valley. Gleser et al. (1981) also reported evidence of sleep disturbances among survivors and of psychological problems among children living in the valley.

In the 1980s there was a marked increase in research on disasters involving toxic waste or radiation. Residents' experiences at Love Canal (a leaking toxic chemical waste dump) appear to have been stressful and should be expected to have chronic consequences for some victims, but the available evidence is inconclusive about this (Levine, 1982). In this case an abandoned short waterway, called Love Canal, in western New York, was used to dispose of 22,000 tonnes of residues from the production of over 200 chemical compounds. This took place between 1942 and 1952. In 1978 it became known that the dump was leaking, exposing the residents to highly toxic material. It was also discovered that women living close to the evacuation site had suffered high miscarriage rates. Similarly, residents in several neighbourhoods in the Netherlands whose houses were built on contaminated soil have experienced stress. In this case substantial proportions of residents continued to feel threatened by the situation because the source of their worries remained (see van der Pligt and de Boer, 1990). Both financial worries, reduced mobility and anxiety about long-term health consequences resulted in a persistently stressful situation. In such circumstances people who have already experience stressful situations tend to be most vulnerable.

The accident at Three Mile Island in 1979 has generated extensive research on both the immediate and longer-term effects on nearby residents. This research has also considered several control groups, has supplemented self-report data with behavioural and biochemical measures and studied residents for over two years. The next section will describe this research.

Consequences of the Three Mile Island accident

The accident at Three Mile Island (TMI) began at about 4.0 a.m. on Wednesday, 28 March 1979, when the TMI nuclear power station outside Harrisburg, Pennsylvania, USA, experienced a major accident. For almost a week radioactive material was released from the plant in an uncontrolled and sporadic manner.

During the accident the reactor core was exposed by a significant drop in levels of coolant water, which generated extremely high temperatures. This heat may have caused a partial meltdown or an explosion of some kind, which fused equipment and resulted in the melting or destruction of radioactive fuel. A considerable amount of water (approximately 1.5 million litres) was contaminated and left on the floor of the reactor building. Radioactive krypton gas was trapped in the containment building surrounding the reactor. Both the contaminated water and the krypton gas became potential sources of radiation; gas occasionally leaked from the containment building into the atmosphere during the year following the accident (it has since been removed). The vast quantity of radioactive water also posed a problem for area residents. The damaged plant was quickly brought under control but remained 'hot' and potentially dangerous for an extensive period of time.

On the day the accident started the local public appears not to have been alarmed. This was partly because many people were unaware until the evening that an accident had occurred. Exceptions to the general lack of concern were those who had close friends or relatives working at TMI. (Those reporting for the 7.0 a.m. shift were not allowed on the island.) On Thursday media reports indicated that the situation was under control. On Friday people began to react to the developments in vastly different ways. Although it was later confirmed that the amount of radiation released was minimal (Kemeny, 1979), information provided by the news media at the time became confusing and was often alarming (see also Goldhaber, Houts and Disabella, 1983). In many areas state policemen went from door to door telling residents to stay indoors, close all windows, and turn off all air-conditioners. All this led to an atmosphere of uncertainty and apprehension. Up to 60 per cent of the population living within 8 km of the reactor site evacuated the area (Flynn, 1979; Hu et al., 1980). The overriding reason given by the people who decided to stay was that they were waiting for an evacuation order. Considering the limited nature of the State Governor's advisory statement (issued on Thursday 29 March), the extent of the evacuation was quite substantial. The advice did *not* *order* a general evacuation but recommended that pregnant women and pre-school children within 8 km of the station should evacuate.

During the emergency period (two weeks), the activities of at least half of the residents in the TMI area were disrupted. For instance, during the week following 30 March, curfews were in effect over much of the area and evening meetings were cancelled. Schools were closed, many of the children were evacuated and, as a consequence, day-time activities involving children were also cancelled. The main changes in day-to-day activities were: staying indoors, cancelling plans, being on edge, and getting ready to leave.

Most residents who left returned within a week, but some stayed away for much longer. Others indicated that they were planning to move permanently from the area because of the threat of TMI (see Flynn, 1979). After about one week the situation at the plant stabilized and within two weeks community life in areas around the plant seemed to be back to normal (Scranton, 1980).

The accident at TMI was one of the first ever serious incidents at a nuclear plant. No one was physically hurt, no property outside the nuclear facility was damaged and, it is generally believed, the accident will not lead to excess deaths or illness (Kemeny, 1979). Nevertheless, as was noted by Goldhaber, Houts and Disabella (1983) psychologically and emotionally the nearby population experienced a disaster. Because of the highly charged issue of nuclear safety, the TMI incident attracted an unprecedented number of investigations of all kinds. Governmental authorities and scientists quickly set up a series of studies to measure the human response to the accident (see Scranton, 1980, for an overview).

Stress and psychological effects

The amount of stress experienced by people around TMI was primarily determined by the perceived amount of threat to physical safety and the perceived reliability of the information about the scale of the threat. The perceived amount of threat varied considerably among individuals. Most respondents thought the threat was very serious (48 per cent) or serious (19 per cent). Only about a fifth (21 per cent) thought it was only somewhat serious, while 12 per cent thought the accident posed no threat at all (Flynn, 1981). Generally, those closer to the plant were more likely to perceive a serious threat than were those further away. Those who thought TMI was a serious threat at the time of the accident were more likely to be younger, female, more highly educated and of high income. Pregnant women were much more likely (at 64 per cent) than average to view it as a very serious threat and much less likely to think it was no threat at all.

Flynn (1981) reported similar results when respondents were asked about their concerns about emissions from the plant. More than 60 per cent of local residents were very concerned with emissions at the time of the accident; 26 per cent were somewhat concerned, while 13 per cent were not at all concerned. Those who did not evacuate were three times as likely to be unconcerned as compared with those who evacuated (19 per cent versus 6 per cent). Considering that the pre-accident evaluations of concern about the TMI plant were moderately favourable, these levels of concern during the accident period represent a substantial change.

Kraybill's study (1979) showed that nearly half the local residents felt that they received insufficient information about emergency procedures. Younger people, the better educated, and those who evacuated were most likely to indicate that they had not received adequate information. Respondents in a study commissioned by the US Nuclear Regulatory Commission (Flynn, 1979) found that media such as local television and radio most useful. Sources such as national network television were regarded less useful, and the print media ranked even lower.

Many studies have discovered psychological effects resulting from the events at TMI, ranging from demoralization, threat perception and fear to increased symptom reporting and negative emotional reactions (see, for example, Davidson, Baum and Collins, 1982; Dohrenwend et al., 1979; Flynn, 1979; Houts et al., 1980). Most studies have reported eventual decreases in these changes, suggesting that problems slowly dissipated after the initial emergency period. Dohrenwend et al. (1979) found that demoralization levels returned to normal within five months. Other studies also reported decreasing differences between TMI and control subjects over time.

Various studies have examined the effects of living near TMI for longer periods of time. These studies suggest that although problems decreased after the accident, they remained at higher levels for as long as two years after the accident. For instance, Bromet (1980) compared young mothers living near TMI to a similar group living near an undamaged nuclear plant. Her findings showed that TMI subjects exhibited increased risk for depression and anxiety and also reported more symptom distress up to nine months after the incident. Houts and Goldhaber (1981) found similar differences in symptom reporting between people living within 8 km of TMI and a control group living further away, up to ten months after the accident. A number of studies compared people living within 8 km of TMI with three different control groups. Results indicated higher levels of stress among the TMI group up to seventeen months after the accident (Baum, Gatchel and Schaeffer, 1983; Collins, Baum and Singer, 1983; Fleming et al., 1982). More specifically, these studies reported greater symptom distress, more behavioural difficulties and increased physiological arousal among TMI area residents as compared with groups living near an undamaged nuclear power plant, a traditional coal-fired plant or no plant at all. Davidson, Baum and Collins (1982) investigated control and stress-related responses of people living near TMI. They argued that one would expect long-term effects from the accident because several sources of stress remained at TMI.

The TMI accident represented a loss of control not only by the residents themselves but also by the authorities: residents were given conflicting information by various authorities which seriously affected the credibility of the information they received. Some people may have feared that they had been exposed to radiation. The inability to do anything to counteract the possible long-term consequences of that exposure and doubts about the authorities' ability to deal adequately with the situation enhanced the feelings of loss of control.

Findings obtained by Davidson, Baum and Collins (1982) indicated that residents reported less perceived ability to control their surroundings (measured by questions about feelings of helplessness, the ability to make meaningful choices, general perceptions of control, and expectations of having control in the future) than did control subjects. Loss of perceived control often leads to cognitive and motivational deficits; this seems to have been the case for residents in the TMI area as well. Davidson, Baum and Collins (1982) measured these deficits by asking subjects to work on a series of eight complex embedded figures similar to the ones used by Witkin, Goodenough and Oltman (1979). Hidden within each figure was a simpler target figure; subjects had to find and trace the correct figure within each item. Experimenters recorded the number of items attempted, the number of correct solutions, and the time spent on each item. Overall, the TMI residents performed worse on the three measured aspects of the embedded figures task (attempts made, number solved, and time spent on the task) than did control subjects. TMI area residents were less successful and also less persistent. Lower persistence and tolerance for frustration have been associated with learned helplessness (cf. Seligman, 1975). The after-effects of uncontrollable noise are similar to those reported above (Glass and Singer, 1972). It is therefore possible that TMI subjects experienced some form of helplessness (see also Chapter 6).

Perceptions of control were also related to the stress experienced by the TMI residents. Because of the stress associated with the accident and its aftermath (i.e., the continued presence of the damaged power plant), TMI residents exhibited greater symptom distress, more behavioural difficulties on tasks and increased physiological arousal than did residents in a comparison group. The findings of Davidson, Baum and Collins (1982) also suggest that feelings of control mediated the stress response. In fact, TMI-area residents who reported lower expectations of controlling their experience also reported greater symptom distress, exhibited poorer task performance and showed higher levels of physiological arousal (as measured by catecholamines) than did TMI residents who reported greater expectations of control.

Support for these findings is provided by Baum, Fleming and Singer, 1982. They began their study when officials were planning to start the controlled venting of the radioactive krypton gas that had been trapped in the containment building. This provided the opportunity to study the anticipatory effects of the venting procedure, the effects of the actual venting and its after-effects. Results indicated higher levels of stress among TMI residents compared with three separate control groups. Moreover, stress levels were highest among TMI residents just before the venting and they continued at a high level compared with the three control groups. During the venting procedure stress levels seemed to decrease somewhat but six months after the completion of the venting they were higher again (Collins, Baum and Singer, 1983). A later study indicated that twenty-eight months after the accident TMI residents continued to exhibit higher levels of stress than people in control areas less affected by the disaster (Davidson et al., 1986).

This study also pointed to the importance of appraisal processes in responses to stressful events. The importance of the appraisal process has been suggested by Lazarus (1966; see also Chapter 6). Davidson et al. (1986) confirmed the mediating role of perceived control in adaptation to stress.

In sum, these studies indicate a persistent stress response – psychological, behavioural and biochemical – more than two years after the accident. TMI residents expressed more concern about possible radiation leaks and more concern about the threat to their personal health than residents of control areas. The studies described in this chapter also suggest that high levels of uncertainty may have exacerbated the loss of control experienced by the respondents and loss of control may have contributed to the continuation of stress experienced by the people living near the reactor site.

What other effects did the accident have on the local communities? As was described in Chapter 1, the accident had a significant and lasting impact upon public opinion, not only in the USA but also in other Western countries. In the next section the effects of the accident on local attitudes towards nuclear power and towards the locality will be briefly discussed.

Other effects

After TMI the attitudes of people in the TMI area towards nuclear power and the restarting of the undamaged TMI Unit 1 became more negative. Attitudinal shifts in the TMI area were similar to those obtained from national opinion polls taken in the same period of time. As was discussed in Chapter 1, these attitudinal shifts became less extreme in the months following the accident but never returned to pre-accident levels.

For some time there continued to be unhappiness about living near the nuclear power plant. When people living around TMI were asked if disliked the area as a result of the accident and wanted to move away, only a small majority wanted to stay. Women, young people and families with pre-school children were most unfavourable towards continuing to live in the area.

Flynn (1979) reports that a total of 19 per cent had considered moving because of the accident (30 per cent of those living within 8 km of the reactor site). All in all 5,100 households within 24 km of the plant (4 per cent) reported that they intended to move. These figures were obtained approximately four months after the accident and further illustrate the seriousness of the local population's concerns. A census conducted by the State Department of Health (Pennsylvania Department of Health, 1979) indicated that 147 households within 8 km of TMI had moved before the end of July (approximately 1 per cent) suggesting that the level of out-migration was modest.

Dohrenwend et al. (1981) showed that the level of public trust in authorities, including federal and state officials and utility companies in the TMI area, was significantly lower than the average obtained in national opinion polls. The level of distrust declined after April, but the decline was gradual. Dohrenwend et al. (1979) report that distrust in the TMI area seems to have remained above the national level in the months following the accident.

Conclusions about Three Mile Island

The accidents at TMI had substantial immediate psychological effects on the people living in the area. The majority of families living within 32 km of the plant evacuated and a substantial minority was demoralized in the month or so after the accident. During the emergency period, the perceived threat, the lack of good communication, the evacuation experience itself, and the psychosomatic effects indicate that part of the population experienced considerable stress. Research also indicates long-term psychological, behavioural and physiological consequences for a (perhaps already vulnerable) minority. The processes of recovery, clean-up and restart all presented additional sources of possible radiation exposure to the general population in the TMI area. Although it seems impossible to predict exactly how the long-term psychological consequences of the accident will interact with the continuing threat, the responses of the local population are bound to be shaped by considerable distrust of the authorities. In a sense, the TMI accident indicates the importance of both adequate risk management and risk communication. The intense public concern following TMI influenced proposals by government and industry for solutions to problems of nuclear power (most notably the risk of a major accident). It has been noted that the journalistic profession might consider adopting a norm of moderation in cases of nuclear accident and attempt to avoid the sensational coverage often provided by unqualified reporters (Mazur, 1984). Bad reporting and the inadequate information given to local residents have certainly accentuated the local response in the immediate aftermath of the accident. These risk management and risk communication issues were extensively studied after another major accident; that at Chernobyl.

The Chernobyl accident

The reactor accident at Chernobyl in Ukraine (then part of the USSR) on 25–26 April 1986 led to the largest release of radioactivity ever recorded in one technological catastrophe. The event is best described as a 'worst case' accident scenario in which a large reactor unit with a

Nuclear accidents: Three Mile Island and Chernobyl

mature fuel inventory (i.e., fully loaded) breached containment and released some of its radionuclide inventory. (All the noble gases, 50 per cent of the iodines, telluriums and cesiums and 3 to 6 per cent of all other materials in the core were released.)

Early in the morning of 26 April 1986 a reactor of 1,000 megawatts in the Chernobyl power station ignited following an explosion. On 5 May Soviet authorities officially announced that the reactor fire had ended and that the reactions between hot steam, graphite and the uranium oxide fuel rods had stopped. For a long time no detailed information was released on the events that led to the explosion and subsequent fire.

According to the USSR Investigations Commission (WHO, 1986), two workers died immediately from the accident, but not from radiation injuries. One died from severe burns, the other when part of the reactor building collapsed. About 200 workers were taken to hospital; it is reported that 18 of these were exposed to such high radiation doses that their condition was severe. Evacuation of the Chernobyl area started on Sunday, 27 April at 2.0 p.m. According to other Soviet reports, some 40,000 people were then evacuated and several days later a further 40,000 were moved out of the area. Later reports mention a total of 130,000 people evacuated after the accident (Renn, 1990).

The first observation of radioactive fallout was made at the radiation monitoring station at Kajaani in Finland, where increased exposure rates were measured on the evening of 27 April. This is consistent with a release at Chernobyl in the night between Friday and Saturday, 25–26 April. Initially it was thought that the radiation might have been caused by a natural radon peak, as had been detected in previous years when snow melted in the spring. However, on Monday 28 April the Rescue Department of the Ministry of the Interior asked for results from its own monitoring stations. Results indicated radiation levels that were 1.2 to 2.5 times the normal values. In Sweden the contamination was first observed at the Forsmark nuclear power station on the Baltic coast, about 100 km north of Stockholm on Monday 28 April. Radioactive deposition was detected in the morning of 28 April, within the site of the station. No reason for the contamination was found within the power station and the Swedish Radiation Protection Institute was alerted.

It was soon concluded that the gamma spectrum of the air activity, which could also be measured at a number of other places along the Swedish east coast, made it unlikely that the activity came from a nuclear explosion. The implication was that there had been an abnormal release from a reactor southeast of Sweden and Finland.

Meteorological trajectories were drawn back towards the Black Sea, though the first guess was that the source was a large nuclear power plant in Latvia. At 9.0 p.m. on Monday 28 April the Soviet news media acknowledged that an accident had occurred at the Chernobyl nuclear power plant.

Between noon and midnight on Saturday 26 April the wind direction shifted. Radioactive material from the accident site released on Sunday 27 April moved west and then south, over the (then) German Democratic Republic, Poland, Czechoslovakia, Hungary, Austria, the southern part of Germany, Switzerland and northern Italy.

In southern Germany heavy rainfall caused increased deposition in the Munich area in the afternoon of Wednesday 30 April. This was consistent with a release on Sunday 27 April, i.e., more than one day after the onset of the accident. Within a few hours the exposure rate increased dramatically and indicated a cesium-137 deposition of about $40\,kBq/m^2$ which is quite remarkable, given the fact that the accumulated deposition of cesium-137 from the atmospheric testing of nuclear weapons was only about $5\,kBq/m^2$ in the 40–50° latitude band.

The total amount of radioactive materials released by the accident was about 10,000 times larger than the emissions from the previous nuclear accident, at Windscale (now called Sellafield) in the UK in 1957. Although the release at Chernobyl was smaller than the total release from atmospheric nuclear weapons testing during the period from 1945 to 1980, the locally deposited concentrations of cesium-137 substantially exceeded the historic fallout from nuclear weapons in countries such as the USSR (now the Commonwealth of Independent States), Poland, Sweden, Italy and Germany.

The initial contamination in Europe affected in Scandinavia and Central-East Europe. Radioactive material released from Chernobyl after Sunday 30 April moved eastwards or southwards, affecting Ukraine, Romania, Bulgaria, Greece, Turkey and the Black Sea region. Contamination found in other countries had essentially been secondary, caused by movements of air masses carrying radioactive material more than five days old (counting from the time of release).

Favourable weather, the relatively remote location of the Chernobyl station, the highly competent Soviet emergency response, the massive evacuation of people and the dispersion of initial releases of radioactive material to high altitude all contributed to a relatively small number of early casualties – only 31 according to a Soviet 'State Committee on the Utilization of Atomic Energy Power Plant and its Consequences' (USSR..., 1986). Air concentrations and deposits as far away as

1,000–2,000 km, however, frequently exceeded levels that demanded protective action from the relevant authorities.

Outside the USSR, radiation levels from the accident were too small to cause any acute radiation effects, though there may be intermediate and long-term biological effects such as cancers and genetic and teratogenic effects. For instance, iodine that enters the thyroid gland increases the probability of thyroid nodules and cancer in this organ, but cancer may not develop for several decades. Teratogenic effects will be evident after birth and genetic effects may appear in one or more generations of offspring. The normal frequency of the various late effects is the result of a variety of influences of which radiation is only one. The additional probability of being affected by some late effect caused by an incremental radiation dose is therefore not easily derived from comparisons with the natural background radiation. As was argued in Chapter 2, it is these combined effects that make epidemiological research assessing the possible health risks extremely complicated.

The distant dispersion of radioactive materials from Chernobyl was dominated by cesium-137 and iodine-131 with some strontium-90. Numerous countries have prepared detailed maps compiling measurements of cesium-137 and iodine-131 (see DOE, 1987; NEA, 1987). Surprisingly, there is no clear 'break' between local dispersion and the distant radiation deposits: in general, the further a country is from Chernobyl, the lower was the average deposition. Significant deviations from the trend, however, did occur and are largely explained by the weather. At a given distance, rainfall produced radioactive depositions 15 to 20 times higher than the depositions found in dry locations. For this reason, several areas in Western Europe (for example, southern Germany, Northwest England and central Austria) received exposures that were comparable to dry deposition within 100 to 200 km from the accident. Some areas in Japan were equivalent to dry locations in Denmark (Wynne, 1989). Similarly, a wet area such as Cumbria (Northwest England) received some of the highest levels of fallout in Western Europe (see Hohenemser, 1988; Wynne, 1989). Coping with such localized 'hot spots' caused confusion to the public and proved to be a difficult task for the relevant authorities.

Initial estimates showed that Chernobyl may impose an incremental cancer risk of less than 0.01 per cent in most locations (Wilson, 1986). However, substantial variations in local exposure occurred which made it necessary for managing institutions to provide different guidelines and to impose different protective actions in different regions. In some countries, the authorities were able to communicate the seriousness of the situation to the public, and also explained why the protective actions undertaken or suggested could differ from one region to another. In other countries there was panic, as in Greece, or distrust of public announcements, as in parts of Germany (see Renn, 1990).

Generally, European governments and relevant agencies were unprepared for an accident of the magnitude and international character of Chernobyl. As a consequence it was necessary to improvise appropriate responses to the fallout. A preliminary survey of actions taken indicates that a wide variety of responses were adopted, ranging from officially planned food-chain protection in the Netherlands to spontaneously generated risk management efforts in South Germany (see Hohenemser et al., 1986). Although overt panic was not reported in the area around Chernobyl, in several countries responses went far beyond the recommendations of the institutions in charge of emergency actions. In Germany and Switzerland, for example, many people removed the upper layer of soil from their garden in order to prevent radioactive material being absorbed by plants and vegetables (Renn, 1990). In other regions, people totally ignored warnings to consume home-grown vegetables.

In the following sections I will briefly discuss a number of issues that played an important role in the aftermath of the accident. In these sections the emphasis will be on aspects different from those discussed in the context of Three Mile Island. The main reason is that most published research after the accident was conducted in locations in Western Europe, and not in the area around Chernobyl. The psychological and behavioural reactions that have been extensively studied around TMI will be more or less absent in the next sections due to the unavailability of published research findings. First I will discuss public reactions; then institutional reactions; I will end with some implications of the accident for risk management and risk communication.

Public reactions to the Chernobyl accident

Opinion polls show that after Chernobyl support for nuclear power declined in most Western European countries while opposition increased. This immediate change was followed by some recovery (see also Chapter 1). But even a year after the accident support had not returned to pre-accident levels. It seems, therefore, that the Chernobyl accident had similar effects on public opinion as did the TMI accident: a dramatic initial increase in opposition followed by moderation but not a complete return to previous levels.

As the previous section indicated, research on stress-related reactions to Chernobyl has been limited. Findings obtained in Sweden (Drottz and Sjöberg, 1990) indicated increased stress in the most affected areas in Sweden. Overall, the effects were less extreme than those at Three Mile Island. Other reports suggest that the combination of health threats and negative consequences to one's livelihood (e.g., as affected farmers in various countries) might have led to stress reactions (see, e.g., Wynne, 1989).

An analysis of newspaper reports by Otway et al. (1987) provides anecdotal evidence of stress-related reactions of the general public in a number of countries. In their study Otway et al. (1987) report a variety of reactions. It often happened that actions taken by people to combat a perceived risk were more dangerous than the risk itself (see also Renn, 1990). Otway et al.'s findings include the following effects of Chernobyl:

- a sudden increase in the number of abortions (reported in both Austria and Italy)
- panic buying of tinned, frozen and other long-life foods, reported in various countries, but reaching near-riot proportions in Greece
- buying of radiation-measuring equipment for personal use (reported in Germany and the UK
- uptake of potassium iodine (sometimes in substantial overdoses), reported in Poland, Germany and Denmark
- an increase in suicides, partly attributed to the inability to cope with the threat, partly attributed to the financial consequences for small farms (reported in Italy and Greece)

Although these extreme reactions received quite extensive press coverage they were not typical of the majority of the population.

Peters et al. (1987) found that 15 per cent of the German population was convinced that the fallout would result in adverse health consequences for themselves; 46 per cent denied that possibility; and 39 per cent were uncertain. A similar percentage (17 per cent) of the families surveyed expected some health damage to their children; 44 per cent did not perceive a health threat; and 39 per cent were not sure. Not surprisingly, those who expected negative health effects for themselves or their children were also likely to be more sensitive to environmental issues and had a more negative attitude towards nuclear power. Of the German respondents 55 per cent declared that they had not changed their diet after Chernobyl. The diet for young children was altered by 54 per cent of the families with children under six years. These changes seem modest given the fact that the German government, along with most interest groups, had recommended changes in diet in almost all areas of Germany (they advised that the consumption of milk, some vegetables and game should be restricted).

The correlation between the frequency of diet changes and fallout dispersion indicated that those who most needed protection did not take more protective measures than those who did not. The changes in diet were clearly related, however, to educational level: the better-educated were more likely to change their diet after Chernobyl. They were even more likely to make changes in children's diets (see Peters et al., 1987). The reluctance of the majority of the German public to take any protective action contrasted sharply with changes in attitudes. More than 70 per cent, including many who did not perceive any threat to their own health, favoured policy options that would ban or at least freeze nuclear energy. So at least initially, Chernobyl left its mark on public attitudes but did not trigger substantial changes in behaviour (Renn, 1990).

In some countries there was a considerable loss of trust and belief in the relevant agencies. In Germany, Sweden and the Netherlands there was sufficient confidence in the agencies handling the emergency. These countries initiated quite extensive information programmes to deal with the aftermath of the accident. However, these countries also saw the provision of often contradictory elements of information by the relevant agencies, industry and the environmental movement. In other countries the picture was radically different. For instance, public opinion polls in Italy and France reported that 70 per cent or more of respondents felt distrust towards, and lack of confidence in, the government (Otway et al., 1987). Before Chernobyl trust in the government was not high in these two countries; their poor handling of information after the accident may simply have aggravated this feeling. France was especially slow in reacting to the accident. It seems as if the authorities initially presumed that the effects of the accident would stop at the French borders. Public concern started when the media reported the effects in neighbouring countries. Only then did the relevant agencies respond. Not surprisingly the initial silence led to increased scepticism and distrust of these agencies.

Institutional reactions to Chernobyl

Most European countries took countermeasures to limit external and internal radiation exposure. The most common measures or recommendations were (see Renn, 1990):

- advice to wash fresh fruits and/or vegetables
- to restrict or ban the import of food from Eastern Europe
- to restrict or forbid grazing of dairy cattle
- to monitor radiation levels in vegetables, fruit and milk
- to restrict the sale of some fresh farm products likely to be affected by the fallout (for example, spinach)

The above measures were taken in most Western European countries, some countries added other measures; for example:

- advice to keep children indoors during rainfall (Austria, Germany)
- advice to avoid drinking rainwater (Austria, Great Britain and others)
- to delay the start of the hunting season (Austria)
- to monitor and restrict fishing (Denmark)
- to close swimming pools, playgrounds and other public places (Germany)
- to remove thyroid glands of cattle (Greece)
- to restrict the sale of sheep (UK)

Due to the extremely limited experience of the relevant agencies, risk communication and risk management were seriously hindered by a veriety of factors. Renn and Hohenemser (1987) mention the following problems:

- difficulties in the design and implementation of a monitoring programme for the detection of local 'hot spots'
- explaining the potential health effects of low radiation
- explaining and justifying different protective actions for different regions or countries
- convincing affected people to follow recommendations and protective guidelines and simultaneously to avoid causing overreactions by those not affected, or only marginally affected
- to be regarded as credible and trustworthy in a situation where most stakeholders in the nuclear debate used the accident to provide the public with their own (often biased) information

As Renn noted (1990), most European countries failed to overcome these problems. They apparently failed in providing optimal protection for the population at risk as well as in assuring the public that a clear, efficient, *and* consistent management approach was taken (see, e.g., Flavin, 1987; Hohenemser and Renn, 1988; Otway et al., 1987;

Wallmann, 1987; Roser, 1987). The confusion was increased by the technical nature of the information, the inconsistent use of units of measurement and the difficulties of conveying the probabilistic nature of the health threat to the public. The 'pragmatic' and sometimes opportunistic use of the issue by specific interest groups (environmentalists versus the nuclear industry) also hindered public understanding. The environmental movement used the situation to argue that all nuclear power stations were unsafe; pro-nuclear agencies focused on the safety of Western-made reactors as opposed to those in Chernobyl and elsewhere in Central and Eastern Europe.

Often citizens were convinced that the government was withholding information and did not tell the truth (63 per cent of the French population, for example). In Germany well-educated citizens complained that the government did not give enough and adequate information, while less educated citizens felt overwhelmed by the flood of information and wished for more consistent and understandable messages (see Peters et al., 1987).

Hohenemser and Renn (1988) argue that since governments were simply not prepared for the international character of the accident, most protective actions involved a good deal of improvisation, inconsistency, and at most a modest amount of prior planning. Decisions concerning protective actions were generally beset by problems. Many countries failed to take or recommend immediate adequate protective action for the population at risk and to assure the public that a clear and consistent risk-management strategy was being implemented. The resulting confusion was further enhanced by inconsistent use of measurement units to indicate radiation levels (rem, rad, sievert, becquerel were all used). Finally, overlapping responsibilities of the authorities also complicated things. For instance, in the Netherlands at least three government departments (Environment, Health and Agriculture) were responsible for the management and implementation of some protective actions. This required the formation of a task force with representatives from all relevant departments and reduced the speed and efficiency of risk-mitigation initiatives.

The major shortcomings of institutional reactions can be summarized under three headings: (1) the setting of allowable radionuclide concentrations in food; (2) the monitoring of radiation levels; and (3) risk communication.

Setting allowable concentrations There were significant national and international differences in the definition of allowable levels of radionuclide concentrations in food (becquerels per litre or kilogram).

Countries varied in the extent to which they based their actions on the same (internationally agreed upon) intervention levels specified in dose units (Hohenemser, 1988). This resulted in substantial inconsistencies. One example from Hohenemser et al. (1986) makes the point. On the West German side of Lake Constance dairy cattle were kept indoors while iodine levels in milk peaked at around 100 becquerels per liter, whereas on the Swiss side of the lake cattle continued to graze on the fresh fallout. The iodine levels in Swiss milk peaked at around 1,000 becquerels per litre. A related problem was created by the failure to integrate the application of national standards with the principle 'as low as reasonably achievable'. Sometimes officials introduced lower levels in accordance with this principle but only after the publication of national standards. This resulted in public confusion due to the presentation of conflicting standards (Renn, 1990).

Monitoring radiation levels Limited experience with monitoring resulted in inadequate assessment and predictions (see also the case study presented in Chapter 8); limited or no attention to local variations; and confusion. This was illustrated by the international trade in food products. Some countries banned food imports from Central and East Europe, even though they did not control their own food supplies.

Risk communication Nearly all countries affected by the fallout had extremely limited experience in communicating specific risks to the general public. Not surprisingly, no risk communication programmes were available. The sensitivity of the issue (radiation) and the fear of the general public about this type of risk required a very careful approach. People confronted with information in terms of 'becquerel per kilogram or litre' in lettuce or milk had no reliable way of translating this information into an idea about health effects and/or comparing the risks with other health hazards. Too often the highly technical information was unintelligeable to the average lay person (see, e.g., Peters et al., 1990; Hohenemser and Renn, 1988; Wynne, 1989).

Improving these shortcomings will require major effort. Some of these improvements are directly related to risk communication issues which will be discussed in Chapter 8. A first requirement is the development of internationally agreed standards for radiation levels that require protective action. Standardized units of measurement are a second requirement. Both should limit inconsistencies in information and improve communication with the general public. Next there should be clearly defined sets of protective actions related to specific radiation levels. These should also help to convey an easily understandable and comprehensible safety and protection strategy to affected citizens. A clear allocation of responsibilities in these emergency situations should help to reach the above three objectives.

Two other issues will be more difficult to resolve: the problem of possible local variations in radiation levels and the public's trust in relevant agencies. The possibility of substantial variations in local exposure to radiation requires monitoring capability at the community or district level. As a consequence, the type of protective action should be tailored to the local situation. Although a single set of protective actions would be most desirable, from the point of view of risk communication, local circumstances and the specific characteristics of the affected region are likely to require different sets of protective actions. Therefore, responses have to be more flexible, and should not be predetermined for each local area. The objective of the flexible response strategy is to accomplish an identical level of public protection using different means. It is clear that such a flexible approach will need considerable effort and substantial financial resources. Some suggestions have been made (see, e.g., Renn, 1990) and these could constitute a first step towards achieving a more adequate and flexible approach. It is more difficult to build trust and credibility. As has been shown, public trust in the relevant authorities varies considerably. Low levels of trust and credibility could hinder effective risk communication, an issue which will be discussed again in Chapter 8.

Conclusions: the consequences of Chernobyl

The accident at Chernobyl had a series of important consequences. Although the number of immediate fatalities was surprisingly low (31), over the next 50 years there may be up to 28,000 delayed fatalities worldwide, about half of them in Ukraine and neighbouring states and half in Europe. These consequences represent approximately 0.02 per cent of the total expected cancer fatalities.

The accident made it essential to improve risk management and emergency planning. A variety of new laws and regulations addressing these issues have been implemented or at least formulated. New provisions for national and local radiation monitoring and food control have seen developed. It is clear that improved risk communication (dealing with both preparation and mitigation of adverse consequences) is an important part of attempts to improve the adequacy of institutional responses to this type of emergency. Risk communication is of crucial importance in accident management. In the next chapter I will turn to the issue of risk communication and provide examples of risk communication strategies after nuclear accidents.

8

Communicating risks

The provision of information about risks takes place in a variety of forms and contexts, ranging from warning labels on consumer products such as cigarettes, to information brochures about health risks, to interactions between government officials and local residents when there have been industrial accidents or releases of toxic waste. Experience has shown that informing the public about risk is much more difficult than it seems. Risk communication on highly complex and charged issues such as nuclear energy has been frustrating for both risk communicators and for those being informed. Before discussing different types of risk communication I will give two examples of failures of public risk communication. The aim of these examples is not to show that most risk communication efforts are inadequate, but to illustrate the problems encountered. One example will deal with disaster warning information at Three Mile Island (TMI), the other with a more specific information programme mounted in the aftermath of the Chernobyl disaster.

Case 1: risk communication after the TMI accident

In their study of the TMI accident, Cutter and Barnes (1982) concluded that local communication links were designed for normal operations. They proved to be insufficient for emergency situations and the resulting blockage of communication channels significantly contributed to the problem (i.e., insufficient information and the resulting uncertainty). With the exception of a direct connection link between the Pennsylvania Emergency Management Agency and the State Police, communication between on-site teams and off-site agencies was handled through a simple manual switchboard which was frequently unattended and could handle only incoming calls (Sorensen and Mileti, 1990). This bottleneck restricted access to the on-site emergency teams and made it difficult to verify information provided by local and county officials. These communication problems contributed to the general sense of confusion and to the ambiguity of information passed to local residents.

Covello, von Winterfeldt and Slovic (1986) refer to another problem emphasized by the TMI crisis: the danger that authorities issue conflicting and contradictory information. Even when they were reporting the same event (e.g., the development of the contaminated gas bubble in the TMI reactor containment) industry spokespersons frequently presented a picture of the risks and dangers very different from that presented by government officials and on-site emergency teams. Statements made by industry spokespersons were generally optimistic and clearly conflicted with messages from the Nuclear Regulatory staff. Four days after the accident began it was decided to centralize the information flow via the Nuclear Regulatory Commission. This alleviated some of the problems. In the meantime, however, the media often presented yet another picture. By aiming for what they considered 'newsworthy', the mass media emphasized information that suggested dramatic events. The resulting confusion was considerable and is seen as one of the causes why so many people evacuated the area against the advice given by the authorities.

Communication at TMI was also affected by mistrust and lack of credibility. Many citizens and reporters felt that there was a credibility problem with both the company running TMI and (to a lesser extent) with representatives of the Nuclear Regulatory Commission. Both were perceived as having an interest in the way the safety of nuclear power was portrayed. As was noted by Covello, von Winterfeldt and Slovic (1986), the motives of the operators of the plant at which a disaster occurs can conflict with those of regulatory agencies and the public. Plant operators could well be more optimistic about their ability to 'fix' a problem than is warranted (Bowonder, 1985). One of the communication problems in technological disasters is that the plant engineers tend to be rather reluctant to release information to people on the outside. They sometimes withhold information about the emerging accident because they hope and believe that they can bring things under control. They also do not want to expose themselves unnecessarily to criticism (see, e.g., Bowonder, 1985; Covello, von Winterfeldt and Slovic, 1986; Lagadec, 1982).

Covello et al. (1990) also noted that nuclear industry officials could have similar biases. For instance, they may emphasize information that avoids damage to the company image and/or reduces the possibilities for liability suits. Both on-site staff and industry representatives could thus be inclined to make optimistic predictions of the consequences of the accidents (e.g., Cutter and Barnes, 1982). This, in turn, could lead to lack of trust and credibility. At TMI it appeared that both residents and reporters most trusted the State Governor and his staff, because they were perceived as having no stake in the causes or consequences of the accident.

This example shows: (1) the importance of adequate communication links in emergencies; (2) the importance of coordinating communication efforts. In emergency situations too much (often conflicting and contradictory) information can have serious consequences (see also Chapter 7). It is possible that the massive evacuation from TMI happened partly because of the ambiguity of the situation, which was created by the substantial discrepancies in information about the seriousness of the accident. There was on the one hand a reluctance to provide information and a generally optimistic 'don't worry' message; on the other hand there was dramatic sensational coverage in the media. The massive impact of confusing communication in emergency situations is also illustrated by some of the events that followed the Chernobyl explosion, as was discussed in Chapter 7.

Case 2: risk communication after Chernobyl

A more detailed case study of what can go wrong at a different stage of emergency risk communication is presented by Wynne (1989), who studied the communication of information to sheep-farmers in the Lake District (Cumbria, UK) during the aftermath of the Chernobyl accident. The main cloud of radioactive contamination from Chernobyl passed over the UK on 2 and 3 May 1986. Virtually no precipitation had disturbed the cloud on its six-day journey until heavy thunderstorms on 2 May deposited radioactive particles across the UK. The Cumbrian fells in Northwest England suffered exceptionally high levels of rainfall – as much as 20 mm in 24 hours.

In the UK rainfall had a clear impact on local deposition of radioactivity (mainly radioactive cesium). It needs to be noted that 1 mm of rain can deposit as much radioactive cesium as 24 hours' dry deposition. In other words, 20 mm of rain in one day can deposit the equivalent of roughly 20 days' dry deposition. Moreover, when rain water falls on uneven terrain, rivulets and puddles can create large differences in radioactivity over even very short distances. Thus actual levels of contamination can vary over small distances and the variability of radioactive deposition does not necessarily correspond with the differences in rainfall. As Wynne (1989) noted in the case of Cumbria, this variability was not fully appreciated at the time.

In the UK 27 monitoring stations provided immediate data for emergency-management. Within a few weeks, a series of about 300 direct measurements of radioactive cesium deposited on vegetation made it possible for scientists at the Institute for Terrestrial Ecology at Merlewood Research Station, Grange-over-Sands, Cumbria, England, to estimate contamination contours in England, Scotland and Wales. The fallout map indicated that Cumbria, especially the southwestern part near Sellafield, received the highest fallout in the UK, over 4,000 becquerels per square metre. Another map of radioactive cesium deposition, produced several months later and calculated from rainfall data rather than deposits on vegetation, produced a different picture: it showed that southwestern Scotland had received over 20,000 becquerels per square metre and the Cumbrian and North Wales fells about 10,000 becquerels per square metre.

Unfortunately these maps were not produced in the first few weeks of confusion and anxiety. In the first weeks, the government issued only bland assurances (cf. Wynne, 1989). A farming journalist described the experience thus: 'The whole two-year event was characterised by great reluctance on the part o MAFF (Ministry of Agriculture, Fisheries and Food) – and indeed other government departments – to inform.... We had to wait for four weeks for the first press release and seven weeks for the first briefing. It was incredible – the silence was deafening' (Oliver, 1988, p. 227).

This overall silence was occasionally interrupted by government statements dismissing the whole event. On 6 May 1986 the Secretary of State for the Environment, Mr Kenneth Baker, assured Parliament that 'the effects of the cloud have already been assessed and none represents a risk to health in the United Kingdom'. Levels of radioactivity were 'nowhere near the levels at which there is any hazard to health'. He stressed that the cloud was moving away and that radioactivity levels falling rapidly. The authorities claimed that there was a steady decline were from insignificant levels of radioactive contamination (see Wynne, 1989).

The British National Radiological Protection Board (NRPB), was more circumspect. On 11 May it stated that the Chernobyl disaster might lead to 'a few tens' of extra cancer deaths in the UK during the next 50 years. This statement was followed by a reassuring forecast that 'if the cloud did not come back the whole thing would be over in a week or ten days.' Baker announced on 13 May that 'the incident may be regarded as over for this country by the end of the week, although its traces will remain' (NRPB, 1986).

The very same day, however, the Ministry of Agriculture, Fisheries and Food (MAFF) found that samples of lamb meat taken from the Cumbrian fells showed contamination levels of over 1,500 becquerels per kilogram. This level was 500 becquerels higher than the European Community's 'action level' (level of radioactivity requiring official intervention). Nevertheless, official statements continued to claim that contamination was falling from already insignificant levels. The Department of the Environment, which coordinated the provision of information, even discontinued its daily bulletins on radioactivity levels on 16 May. The argument was that levels were now insignificant.

On 30 May MAFF announced 'higher readings' of radioactive cesium in hill sheep and lambs. The Ministry concluded that 'these levels do not warrant any specific action at present', on the assumptions that hill lambs were not yet ready for sale and that the high levels of contamination would soon decrease (see Wynne, 1989 for a more detailed account). On 20 June, however, the Secretary of State for Agriculture announced an immediate ban on both the movement and slaughter of sheep in specific parts of Cumbria and North Wales. This advice constituted a remarkable change from the previous statements. Even this totally unexpected intervention was laced with reassurances that the ban would have minor consequences because radiation would fall to acceptable levels before lambs were ready for market. The ban was imposed for only three weeks, based on the assumption of declining radiation (Wynne, 1989). Unfortunately, levels of radiation in sheep increased rather than decreased. On 24 July the ban in Cumbria was extended indefinitely.

The areas restricted on 20 June contained over 4 million sheep. Four days later, parts of Scotland were added to the restricted region. By the end of September 1986 the restricted region in Cumbria had been reduced from 1,670 farms to 150 farms. The latter 150 farms *remained* restricted. In June 1987 some farms in North Wales and Scotland that had been freed of restriction were rerestricted. Moreover, some farms in Scotland and Northern Ireland were restricted for the first time. Overall about 800 farms in the UK with over 1 million sheep were still restricted in March 1988, nearly two years after the accident!

This episode shows an extremely erratic pattern in the provision of information. The 'experts' first issued reassuring statements claiming that there would be no problem; then that unexpected restrictions would last only a few weeks. Later the same experts declined to predict how long the restrictions might continue (HMSO, 1988). Not surprisingly, this lack of consistency was utterly confusing for the farmers involved and resulted in considerable uncertainty for most of them.

To make things worse, the resulting lack of trust and credibility was accentuated by rumours that the Sellafield nuclear reprocessing plant in Cumbria might *also* have contributed to the contamination. Because the claimed contamination from Chernobyl occurred near Sellafield the suspicion arose that contamination of the Cumbrian fells had always been high. Sellafield's routine discharges and/or the 1957 accident were seen as possible causes. It was possible that radioactive deposition simply had not been monitored or admitted before the Chernobyl accident. Wynne (1989) noted that such concerns were widely expressed by residents after the Chernobyl accident. Experts did not agree and stated that the isotope ratio of cesium-137 and 134 in Sellafield discharges could easily have been distinguished from the fresher Chernobyl deposits. Because the half-life of cesium-137 is about 30 times that of cesium-134, the ratio of cesium-137 to 134 increases with time. The Chernobyl fallout ratio was about 2 or 3 to 1, the ratio of the remaining isotopes from the 1957 Sellafield fire was much higher, about 12 to 1. This enabled experts to identify the source of the radioactive fallout.

Not surprisingly, farmers were unconvinced by these analyses. Interviews with farmers revealed the widespread belief that contamination from Sellafield had contributed to the high levels of contamination and that it had existed unmonitored for a considerable length of time. The farmers had been asked to believe analyses of contamination by the same experts who had recently shown themselves to be equally confident (but totally wrong) about the rate of decline of the contamination caused by Chernobyl. As Wynne (1989) notes, this false certainty of the scientists was frequently cited as a cause of their lack of credibility. When on 30 September 1986 the initial ban over nearly the whole area was reduced to a much smaller area around the Sellafield plant, farmers became even more confident in their judgement that background radiation from Sellafield was a major cause of increased levels of contamination.

Wynne argues that the farmers' scepticism about Sellafield's claimed innocence was consolidated when local requests for radioactive monitoring data collected before Chernobyl were effectively turned down to avoid the embarrassing acknowledgement that hardly any such data had been gathered. Not surprisingly, the farmers drew their own conclusions about expert credibility. As a consequence, government technical officials were generally seen to be incompetent. Whether research after the 1957 fire had been overlooked or was being covered up, neither explanation benefited the experts' credibility. This was all vindicated by a survey published in March 1988 which argued that half the contamination on the Cumbrian fells was from Chernobyl and half was from Sellafield and from nuclear weapons' testing fallout combined (cf. Wynne, 1989).

This case study demonstrates the importance of ensuring consistency in information and the effects of a lack of it on the trust and credibility of the relevant authorities. Entirely separate issues – the 1957 Windscale fire, routine discharges from Sellafield and the Chernobyl accident – became more or less the same issue to local residents and further damaged the credibility of the various authorities. The numerous, and not always understandable, revisions in the restrictions enhanced this problem. Unfortunately this inconsistency spread into the rules devised for the allocation of economic compensation. By 1989 only some farmers had received compensation for financial losses that resulted from the restrictions (cf. Wynne, 1989).

In this episode central information sources became distrusted and lost most of their credibility. Too often the information was provided too late. It was followed by conflicting evidence a few weeks later. Information usually seemed devised to focus on short-term goals (e.g., quieting down local protest or concern). Wynne (1989) argues that the achievement of such short-term goals tends to be attained at the cost of long-term disorientation and instability in the relationship between the public and the authorities and their experts.

In the following sections I will consider risk communication at a more general level and discuss several types of risk communication tasks. Three types of risk communication tasks are of relevance to the present chapter: (1) general information and education; (2) hazard awareness and emergency information; (3) policy decision-making and conflict resolution. As Covello, von Winterfeldt and Slovic (1986) noted, these tasks frequently overlap, but they can be conceptually differentiated.

Information and education

The provision of information and education about nuclear risks generally takes place in two specific circumstances: (1) after a nuclear accident (e.g., Sellafield, TMI, Chernobyl); (2) in the process of siting nuclear waste disposal sites or nuclear power stations. In both cases the provision of information can take several forms, ranging from brochures and mass media attention to public hearings. This section will focus on information programmes that attempt to provide general information about radiation, risk of exposure to radiation and the possible adverse consequences of exposure. Some of the problems that complicate the task of informing and educating people about risks have been mentioned (implicitly or explicitly) in the previous chapters. The major problems can be summarized under the following headings:

Complexity Information about risk is often highly technical and complex. For the average lay person quantitative risk information is difficult to comprehend and relatively meaningless. Complexity is further increased by the need (sometimes the habit) to convey this sort of information in scientific, legalistic and formal language. The combination of these factors can cause the public to believe that risk communication efforts are evasive and not to the point. The public is left feeling confused and suspicious.

Uncertainty Limited experience often leads to substantial uncertainties on the part of experts. This can be caused by inadequate data bases and the shortcomings of available methods and models. Experts have disagreed about the validity of risk assessments. They often disagree about the assumptions and procedural aspects of risk assessment techniques and as a result often provide widely differing estimates (see also Chapter 2).

Changing perspectives Improvements of both data bases and assessment techniques generally result in more reliable risk assessment. Rapid changes in assessments, however, also create discontinuity and confusion for the public. Such consequences were illustrated by the frequent re-assessments and changing policy measures in Cumbria that were described earlier in this chapter.

Frame of reference Lay people and experts often use different frames of reference when evaluating risks. As a consequence they use different definitions of risks (see Chapter 2). Lay people tend to focus on factors such as catastrophic potential, fairness of the distribution of risks and benefits, effects on future generations, voluntariness and controllability. Experts, on the other hand, tend to define risks in terms of expected annual mortalities. These different frames often result in a reluctance to accept the risk management solutions of governmental agencies.

Trust and credibility Governmental agencies and industry sometimes lack public credibility and trust. Recent history has shown that their limited experience of being accountable to the general public has sometimes led to the provision of limited information and/or deliberate withholding of information (usually because agencies, the industry and other relevant bodies have feared emotional reactions and/or panic). This, together with perceptions of technical incompetence, biases in favour of specific interest groups and the presumed excessive weight given to economic factors have severely damaged the public trust and the credibility of some governmental agencies and the nuclear industry.

Involvement and concern Sometimes it is difficult to predict public involvement with, and concern about, specific risk issues. On the one hand, people can be far *less* interested in risks than governmental agencies. In these cases it is difficult to make people take risk information seriously (for example, information about the risks of smoking, alcohol or diet). On the other hand, people can be far *more* interested in risks than governmental agencies. In these highly charged areas (e.g., the dangers from contaminated soil in residential areas, industrial accidents, nuclear radiation leaks) risk communication is particularly hard and needs very different strategies from situations in which one (desperately) tries to attract the attention of the public for less salient risks.

These six problem areas show the difficulties of risk communication. The first three are related to the characteristics of the issue, the fourth and sixth to differences in conceptualization between lay people and experts. Both frames of reference and involvement need to be assessed more carefully before risk communication efforts take place. The fifth factor (trust and credibility) refers to the relevant agencies' lack of experience, their underestimation of the public, and consequently, their relatively poor history of communication.

In the last part of this chapter I will deal with ways to improve the performance on these factors, but before this is considered another type of risk communication requires discussion: the provision of emergency information.

Emergency information

Hazard awareness programmes provide information and education for the public about natural and technological hazards. They may cover the nature of the hazard, ways of mitigating the risks, and protective actions to be taken in an emergency. Emergency warning systems provide information to the public when an event is about to occur or is taking place. During the 1980s researchers have produced a substantial literature on disaster communication, relating to such disasters and emergencies as the TMI accident (Cutter and Barnes, 1982), the Bhopal accident (Bowonder, 1985), and the accident at Chernobyl (Renn, 1990).

This section is organized in two parts. The first deals with communications aimed at educating the public about possible disasters and preparedness for an emergency. Research in this area has concentrated upon the effectiveness of risk communication programmes at raising levels of public awareness and preparedness. The second part deals with disaster warning systems and the communication of information in the course of a disaster.

Emergency preparedness Covello, von Winterfeldt and Slovic (1986) note the paradox that disasters make dramatic headlines and tend to dominate public opinion *after* they occur, whereas it is extremely difficult to interest the public in an issue *before* the immediate threat of a serious accident. The research literature on natural disasters indicates that people are generally not interested in disaster information and rarely make plans for possible emergencies (see, e.g., Flynn, 1979; Turner et al., 1981).

One possible explanation for this lack of interest is what Weinstein (1980) called 'unrealistic optimism'; i.e., the belief that a disaster is more likely to happen to other people than to oneself. This optimism is also found in the context of technological disasters. For example, people tend to be overly optimistic about the safety of dams and they are reluctant to buy disaster insurance (Kunreuther et al., 1978). Similarly, van der Pligt, Eiser and Spears (1987b) found that people living near a nuclear power station thought that the risks of accident and routine emissions applied less to their local power station than to similar facilities elsewhere.

Research on technological disasters indicates that the general public overestimates the probabilities and consequences of rare technological disasters (see also Chapter 2). Moreover, people are more pessimistic about the likelihood of a nuclear power accident than even the most pessimistic expert (see, e.g., Slovic, Lichtenstein and Fischhoff, 1979; von Winterfeldt, John and Borcherding, 1981). One might expect that people who hold a pessimistic view of the possibility of a disaster would develop plans for an emergency, yet there is little evidence that this pessimism is translated into improved planning for disasters and emergencies. Covello, von Winterfeldt and Slovic (1986) argue that this could be related to uncertainty fuelled by expert disagreement about what constitutes an effective emergency response. On the other hand, it must be noted that quite often the relevant agencies are hardly prepared themselves and seem to share the optimism we referred to earlier. Examples described in this chapter and also the aftermath of the Chernobyl disaster (Chapter 7) illustrate the limited attention paid to possible disasters.

Several studies have looked at the relationships between awareness of risk, judgement of risk and proximity to a possible source of disaster. Several studies (e.g., Maderthaner, Guttman and Otway, 1978) found an inverse relationship between distance from nuclear power plants and perceived risk; i.e., the nearer a person lives to a plant, the higher the perceived risk. It needs to be added, however, that this relationship does not seem to hold for people who live very close to a plant (within 5 km). As has been stated before, people who live near a nuclear plant and are presumably subjected to the greatest risk do not perceive the risk as great (see also Chapters 2 and 4).

Several suggestions have been made to address the limited public awareness of disasters and the lack of preparendness for emergencies. Covello, von Winterfeldt and Slovic (1986) list several recommendations for the development and implementation of educational programmes aimed at improving public knowledge about disasters and preparing them for disaster contingencies. These can be summarized as follows: (1) increase involvement; (2) avoid panic and excessive concern; (3) be as concrete as possible. Unfortunately, little is known about how to achieve these aims. The latter is not an easy task, and only in the 1980s have field experiments addressed these issues, some successfully, others with mixed success (see also Earle and Cvetkovich, 1985).

Most studies (see, e.g., Sorensen and Mileti, 1990) illustrate the need for there to be several channels of communication in hazard information programmes. People rely on, and pay attention to, different sources of information. Specialized programmes may be more effective if they are linked with the general media (television, radio, newspapers) since people use these most often as sources of information.

Little is known about why pre-emergency risk communication programmes produce different awarenesses of risk and knowledge about what to do in an emergency situation. Baumann (1983) found that the amount and intensity of information had no effect on knowledge commumicated. Neither the source of information nor its judged usefulness determined the public's knowledge about what to do in an emergency (see Sorensen, 1983). McKay (1984) concluded that personalized information (for example, information presented by a safety executive visiting the neighbourhood) is more effective at increasing knowledge.

Sorensen and Mileti (1990) put forward the view that there is no conclusive evidence that people are more prepared and protected as a result of information programmes. Unfortunately, most research about how programmes could be improved to produce higher levels of knowledge and protection is inconclusive. The experience gained with one programme at a particular location may not always be useful in designing programmes for different locations or for an entire country.

Disaster warnings The difficulties encountered in general disaster education and preparedness for emergencies (e.g., limited public interest and awareness, unrealistic optimism and a lack of planning for disasters) are reflected in the difficulties encountered in designing and operating pre-disaster warning systems. These systems should rapidly attract people's attention, alert people to impending danger, instruct people to be prepared for further action and, finally, to communicate these actions clearly.

Evacuation and sheltering are the commonest and most effective means for mitigating the consequences of disaster. The literature on evacuation programmes indicates that in many natural disasters people are reluctant to evacuate or use shelters. Covello, von Winterfeldt and Slovic (1986) suggest that both the reluctance to evacuate and to use public shelters (people tend to prefer homes of friends) are the result of fear of losing control over one's belongings and fear of looting (see also Chapter 7).

It has been argued that, in contrast to natural disasters, technological disasters produce substantial compliance with evacuation recommendations – in some cases even excessive evacuation (Flynn, 1979; see also Chapter 7). Studies at TMI have shown that many more people (about 145,000) evacuated than were advised to do so (Cutter and Barnes, 1982; see also Chapter 7). Studies that presented people with a hypothetical nuclear power plant accident also indicate 'excessive' evacuation intentions (Covello, von Winterfeldt and Slovic, 1986). Similar findings were obtained at Love Canal (Levine, 1982).

Cutter and Barnes (1982) studied evacuation behaviour during the Three Mile Island accident. They found that approximately 40 per cent of those living in the vicinity of TMI evacuated after it was recommended that pregnant women and pre-school-age children living within 8 km of the reactor should evacuate and that people living in a within 16 km zone should stay indoors and keep their doors and windows shut. Nearly all evacuees moved to private accommodation, either their own or that of friends. Only a small majority used an official evacuation centre (Flynn, 1982). Cutter and Barnes also noted that elderly people were much less likely to evacuate than younger people. At TMI only about 15 per cent of the elderly evacuated among those living within 16 km of the plant, while about 50 per cent of the younger people evacuated. Reasons for remaining included health problems and reluctance to leave one's home for fear of losing property, valuables and contact with friends. Younger people who decided not to evacuate said that they stayed because they were waiting for further and stronger evacuation orders. They did not feel threatened but feared looting.

There was considerable confusion at TMI as a result of conflicting and sometimes even contradictory information. As was argued in the introduction of this chapter, confusion was one of the major causes of the observed excessive evacuation. Another important finding was that people usually do not react to a single message or instruction from a single source. Generally, people will search for confirmation from other sources (see, e.g., Cutter and Barnes, 1982).

While it is important for a single authority to issue initial announcements about necessary emergency actions, this information should be broadcast via television and radio and, in more urgent situations, also through mobile emergency teams using loudspeakers. Obviously, reliance on a single system, such as sirens, tends to be less effective, since many people forget what these signals mean (see also Hodler, 1980).

An important aspect of emergency information systems is the availability of communication channels. Communication links designed for normal operations tend to be insufficient in disaster situations. The resulting bottlenecks can contribute to the general sense of confusion. Quite often this is enhanced by the inconsistency of information provided by different sources and by the media selecting news on the basis of 'newsworthyness'. Centralizing the information flow helps, but is not always sufficient. For instance, Sandman and Paden (1979) found that many people in the vicinity of the TMI plant received most information indirectly, and after considerable delays, via radio and television reports.

Although there was no evidence of panic behaviour at TMI, rumours that occur when information is contradictory and ambiguous could result in overreaction and panic. The excessive evacuation at TMI and some reactions after the Chernobyl disaster (see Chapter 7) can be related to rumours. Perry, Greene and Lindell (1980) recommend the use of telephone 'hotlines' to enable people to get confirmation or denial of specific rumours. This strategy was successfully adopted in the Netherlands after the Chernobyl disaster. A national hotline was established shortly after the accident. Establishing such a hotline in an actual disaster area, where it is most needed, is however much more difficult, as the TMI accident showed.

The major lessons drawn from disaster experiences can be summarized as follows:

(1) a communications network adequate for emergencies is essential
(2) information should be provided by a single source using multiple communication channels (printed media, radio, television and mobile teams with loudspeakers)
(3) give precise, detailed information about evacuation and/or shelter
(4) establish a central telephone service for providing information to the public on request

Policy decision-making and conflict resolution

Since the mid 1970s public involvement in the nuclear issue has increased dramatically. People have come to expect a right to know and to participate in policy decision-making that affects their lives. For instance, in Europe, the Seveso accident had far-reaching implications for the European Community in the form of the Seveso Directive, which imposes obligations for both the assessment of technical risk and its communication to the authorities and the public. (The accident happened in 1976 in a chemical factory near Seveso in North Italy and resulted in a release of highly toxic dioxin.) In the United States the 1987 Superfund amendments expanded responsibilities for hazard communication by making the provision of information about specific hazards and risk mitigation obligatory and substantially increased local participation in hazard-related assessment issues. From 1975 onwards various initiatives have been taken to try and involve citizens more directly in the formulation of nuclear policy. One example is the public inquiry.

Public inquiries play a role in public policy decision-making and focus on specific projects (e.g., the development of a nuclear facility). They also provide the opportunity to express a range of opinions. In most cases, the agenda is limited and the proceedings tend to be dominated by scientists. Inquiries tend to be guided by the assumption that factual evidence should carry more weight than subjective, more often local, concerns.

This was the case in the public inquiry into the plan for THORP, a thermal oxide fuel reprocessing plant at Sellafield in the UK (see also Chapter 3). Many organizations testified at this 100-day inquiry which aimed to assess conflicting arguments in order to evaluate the proposed development. With no source of funding, objectors had great difficulties in developing a coherent and coordinated position to counter the arguments of British Nuclear Fuels Ltd. Similar inquiries took place about the proposed development of Sizewell (the Sizewell B nuclear reactor) and Hinkley Point (a third nuclear power station at Hinkley Point in Southwest England). These inquiries took considerably longer than the Sellafield inquiry (Sizewell over a year) or were (temporarily) halted (Hinkley Point) because of the uncertainty about Britain's nuclear future.

In the Netherlands the government organized an extensive consultation of the public through a massive public public in the early 1980s. It was based on the principle that the public should be consulted on overall energy policy decisions. A more cynical interpretation of this policy is that the government's own indecision was transferred to the public by asking it to discuss possible energy futures. Government plans were accompanied by the publication of 'policy intentions' dealing with issues such as the objectives of economic growth, the goals of particular energy scenarios and their likely impacts (see Nelkin and Pollak, 1977; Vlek, 1986). These were widely distributed for public comment. The entire dossier developed through this process served as a basis for a Parliamentary discussion. Unfortunately, shortly after this inquiry, which revealed that a majority of the public was opposed to expanding the number of nuclear power plants, the Department of Economics announced plans to develop two more nuclear power plants in the Netherlands. The announcement damaged the credibility of the public inquiry system. The Chernobyl accident then forced the Dutch government to postpone decisions on the expansion of nuclear power.

All too often, public inquiries have seemed to be essentially cosmetic inquests into public objections and complaints about specific projects. Nelkin and Pollak (1977) argue that this is the purpose of the French Déclaration d'Utilité Publique (DUP) and certain provisions of the German Atom Law. To build a nuclear power plant in France, Electricité de France (EDF) must undertake an inquiry through a DUP procedure (Colson, 1977). The DUP has become a forum for the expression of concerns about environmental risks, health issues and personal safety. In the DUP process, all people living within 5 km of a proposed nuclear facility have access to a technical dossier for 6–8 weeks, during which time they can formulate their objections. The local prefect appoints an investigating commissioner (usually a local dignitary) who collects and evaluates the complaints and the EDF's response. On the basis of this evidence it is recommended whether the project should be given public utility status. Up to now, no commissioner has ever rejected an EDF application (cf. Nelkin and Pollak, 1980).

Under the German Atom Law, electricity companies wishing to build a nuclear plant must apply to the Land (State) administration for a construction permit and hold a public hearing. Documents are available for public inspection for a month. Anyone affected by the project is entitled to object – the courts have accepted claims from people up to 100 km from the plant site (see Nelkin and Pollak, 1980). Generally, this hearing is restricted to the nuclear safety issue; other environmental issues and economic and social concerns are considered irrelevant. As Nelkin and Pollak (1980) noted, this restriction frequently limits meaningful participation by lay people.

Some countries have developed other initiatives to increase public participation in nuclear energy decision-making. For instance, Sweden initiated an experiment in public education about energy issues. The experiment included a system of citizen study groups managed by a variety of organizations and political parties. The government financed a programme to inform broad segments of the public about energy and nuclear power (Nelkin and Pollak, 1977). The programme involved some 8,000 study circles, each with about ten members, who met to discuss energy-related questions they felt to be most important. The expectation was that greater information would create more favourable attitudes towards government policy. Reports from these groups, however, suggest continued uncertainty and ambivalence (Nelkin and Pollak, 1980).

The *Bürgerdialog* in Germany represents a similar effort to involve broad sectors of the public in a risk information programme. Organizations such as churches, unions and adult education groups are funded organize discussion groups and meetings. These include speakers both for and against nuclear energy. The goal is 'to strengthen confidence in the ability of the democratic process to function, especially in the controversy over nuclear energy, and to restore confidence wherever it may be undermined' (Nelkin and Pollak, (1980)). The primary aim of this effort is to inform citizens about the necessity for nuclear energy and convince them that the risk is minimal. But as Nelkin and Pollak (1980) pointed out, the 'dialogue' has frequently become a monologue, and it has clearly failed to create consensus over nuclear policy.

Too often the outcomes of these initiatives suggest that there is not really a choice. After the national debate in the Netherlands a clear majority of the public opposed any expansion of the nuclear industry, yet shortly after the publication of the final report the Department of Economics announced that it would ignore the outcome of the inquiry. Nelkin and Pollak (1980) mention more extreme examples of authorities ignoring the outcomes of public inquiries. For instance, in 1977 a public inquiry for a nuclear plant opened at Le Pellerin in France, where there had long been local resistance. The mayors of seven of the twelve communities in the region had refused to use their offices for the inquiry. The population agreed to boycott the official inquiry. Of the few people who did participate, 95 approved the project, 750 opposed it. In all 80–95 per cent of the population in each of several communities signed anti-nuclear petitions. The outcome of the inquiry seemed to ignore all of this: the inquiry commissioners declared themselves incompetent to judge the issue, but concluded in favour of the project.

This is an extreme example, but it suggests a low tolerance for disagreement in public inquiry procedures. Most public inquiries have looked like structured discussions over predetermined policy. The financial and administrative investments involved in specific technologies are generally to massive to allow a real margin of choice. In the case of a plant in Esensham, Germany, regional officials acknowledged that secret negotiations with the nuclear industry had taken place for over a year before the industry lodged an official application for a construction licence. In several cases Electricité de France began preparatory construction work before the end of inquiry procedures (see Nelkin and Pollak, 1980). In the UK the public inquiries into THORP at Sellafield and the building of a nuclear power plant at Sizewell left the opposition with mixed feelings. The unequal distribution of expertise and financial resources and the limited attention paid to views of the nuclear opposition made opponents quite cynical about the inquiry procedures. The perceived lack of a real choice results in the view that public inquiry systems should be used primarily to delay (foregone) conclusions. For instance, the public inquiry about Sizewell took over 350 days; the final decision favoured the building of a new nuclear plant.

After initial decisions have been made and, for instance, shortlists of possible locations have been decided upon, risk communication enters a different stage. Quite often, risk communication tasks involve direct personal communication between representatives of public groups and officials of the responsible authorities with the objective of formulating policy decisions and/or resolving conflicts. In the 1980s public participation at a local level increased, partly because local acceptance became a precondition for solving urgent problems such as the siting of nuclear waste facilities. Examples of this kind of public participation include communications between public officials and local citizens at community meetings and communications between public officials and representatives from interest groups. Not surprisingly, these interactive communications frequently involve highly emotional confrontations between opposing groups. Stakeholder groups usually include proponents of the technology, opponents and government officials.

The involvement of the public in this stage of policy decision-making leads to numerous communications difficulties Kasperson (1986) identified several factors that have limited the success of local participation efforts:

(1) Conflicts are often related to differences in expectation. Local residents often see risk communication programmes as an invitation to share power, to participate in *defining* ends. The responsible authorities, on the other hand, will see participation as having primarily an *instrumental* function, as a means to find acceptance for specific policy decisions.
(2) Lack of involvement in the early stages of decision making: quite often, public participation is limited to the later stages in risk management issues. Major choices have already been made and other options are foreclosed. Both the developer and governmental agencies are committed to programmes. In these cases little opportunity exists for the public to influence the decision process.
(3) The acceptance of information about the risks of a nuclear facility is clearly related to an institution's credibility. Public participation requires trust that the public's views will be taken seriously in the ultimate decision.
(4) Effective public participation depends largely on the availability of resources and the means to act on increased knowledge. Participation in the absence of resources tends to be ineffective and tends to reduce the credibility of participation programmes.
(5) Disagreements about local developments tend to be quite complex and involve both disagreements about facts and disagreements about values. Failure to disentangle these often hampers effective risk communication.

The latter issue is especially important for risk communication in the context of policy decision-making and conflict resolution. Disagreements about facts are most often disagreements about (1) data, (2) risk assessments, and (3) assumptions and definitions (see also Chapter 2).

Disagreements about data usually concern the reliability and validity of obtained findings (e.g., epidemiological research on relatively small samples). Disagreements about risk assessments and probabilities tend to be the result of different conclusions reached by different experts. Examples of disagreements about risk estimates include the probability of meltdown at a nuclear power plant and the long-term health consequences of low-level radiation. Finally, disagreements about assumptions and definitions refer possible models for characterizing risk and/or the definition of risks. Examples of disagreements about assumptions and definitions are disagreements about how to weigh delayed deaths and other health effects in risk assessment, what health effects to include in risk assessments, and the relevance of qualitative considerations related to psychological risks (e.g., voluntariness and controllability) in characterizing the risk of a nuclear power plant.

As we saw in Chapter 2, disagreements about values usually concern disagreements about (1) risk–benefit tradeoffs and distributions, and (2) basic social values. Disagreements about risk–benefit tradeoffs most often arise when one group stresses the risks of a technology (health hazards, accident potential) and questions its benefits (benefits for consumers, for the economy) while another group reverses this emphasis. Examples of disagreements about risk–benefit tradeoffs and distributions are conflicts over locally unwanted developments such as waste facilities (see also Chapter 5).

Disagreements about basic social values are related to different conceptions about society as a whole and they constitute a more fundamental level of conflict (see, e.g., Nelkin, 1979). Examples include debates about nuclear power centred on issues such as economic growth and the fairness of transferring the nuclear waste problem to future generations. The above discussion makes it clear that value-issues play an important role in the nuclear debate. Moreover, they will tend to dominate risk communication efforts in the context of policy decision-making and conflict resolution. Risk communication alone will not solve these value conflicts. One solution to this problem is to adopt a decision analytic approach: to analyse conflicts, disentangle the disagreements about facts from disagreements about values and create options that help address the values relevant to all the stakeholder groups. Chapter 9 will discuss decision analytical techniques in more detail.

Conclusions

In the introduction of this chapter two case studies served to illustrate the wide variety of problems encountered in informing the public about emergencies. The lessons learned from the events at TMI and Chernobyl point to the importance of having adequate communication networks and to communicate via a single source of information using a variety of communication channels. The latter arrangement should help to reduce inconsistencies. Finally, it is advisable to establish means for the public to check rumours. This could be done via a central telephone service.

In the remainder of the chapter I distinguished different purposes of risk communication. The first is to inform and educate the public. This general aim is more difficult than it seems. The major reasons for the limited success of nuclear risk information programmes are related to the specific characteristics of the risks involved, the different frames of reference of expert communicators and the lay public, and the rather unfortunate history of communication attempts which eroded public trust and damaged the credibility of the industry and responsible agencies.

The second purpose of risk communication concerned the improvement of hazard preparedness and the effectiveness of accident warning systems. The success of programmes aiming to increase preparedness for possible nuclear disasters seems to be limited. For this the major reason appears to be the low involvement of the public and their limited willingness to think about their preparedness for possible nuclear disasters. Both conflicting information (if any is provided at all) and the optimistic view that major accidents happen elsewhere seem responsible for the limited impact of communication efforts aiming to increase the general hazard awareness and preparedness. Accident warning systems have been studied more extensively. The first case study described in this chapter dealt with that issue.

Finally, risk communication could also help formulate policy solutions and could play an important role in the resolution of conflicts about nuclear energy. The 1970s and 1980s saw several attempts to involve the public in nuclear energy decision-making. Unfortunately these attempts met with only limited success, the major reasons being the different perspectives of the various stakeholders and the unfortunate history of inadequate attempts to involve the public in conflicts about nuclear energy policy.

Risk communication thus involves a variety of complex tasks. In the late 1980s some attempts were made to translate risk communication research into practical risk communication advice. Most prominent among these efforts were guidebooks aimed at the representatives of government agencies (e.g., Hance, Chess and Sandman, 1987) and at managers of hazardous materials plants (Covello, Sandman and Slovic,

1988). A related effort is a publication by Covello and Allen (1988) called *Seven Cardinal Rules of Risk Communication*. These efforts have met with some criticism. One reason is because the manuals tend to give the impression that much useful information is available about risk communication processes. However, this is not the case. Moreover, much information about risk communication concerns a specific hazard and cannot always be applied to other hazards. The need to be of use in a variety of situations and contexts has resulted in lists of very general recommendations described by some as hollow platitudes (Earle and Cvetkovich, 1988). Advice such as 'accept and involve the public as a legitimate partner' (Covello and Allen, 1988) is not particularly helpful unless the advice deals with ways to do this successfully. It could well be that these first efforts to translate risk communication findings into practical manuals were made too soon, but they also illustrate the pressing need for practical advice.

Some implications can be drawn from the available research as we have seen in the present chapter. Key concepts that apply to the whole spectrum of risk communication are 'consistency' and 'perspective'. Lack of consistency has done much damage in the past and will continue to do so. Perspective refers to a variety of differences between experts, communicators and lay people. These include different definitions of risk but also different views on cost–benefit issues. Tools to help disentangle differences of opinion about facts and values between the various groups seems a prerequisite for successful risk communication. Chapter 9 will discuss some of these tools.

9
Decision analysis and nuclear energy policy

Von Winterfeldt and Edwards (1986) have argued that decision analysis is still an emerging discipline and that it is more often 'sold' by suppliers rather than sought by customers. This picture is changing rapidly and, as we will see later, some decision-aiding techniques have been used for nuclear energy decision-making. Decision analysis consists of a variety of tools that could help decision-making. These tools are related to various stages of the decision-making process: some are designed to help structure and define a problem; others focus on the elicitation of probabilities and utilities; some apply to models designed to help evaluate alternatives.

Decision analysis provides a set of formal models and semi-formal techniques to help structure and simplify the task of making complex decisions. Decision analysis techniques thus seem most appropriate in the context of complex, confusing and sometimes stressful decisions. As we saw in the previous chapter, nuclear energy policy decisions frequently fall in all of these categories.

Decision analytic tools are particularly helpful in two complex areas of decision-making: uncertainty and multiple objectives (cf. von Winterfeldt and Edwards, 1986). One of the prevailing issues in nuclear energy policy is uncertainty about risks and future developments. Two techniques focus on this uncertainty: scenario construction and risk assessment techniques. Decision analysis can also be helpful when there are multiple conflicting objectives. Almost all important decisions incorporate multiple values. Sometimes these go together, sometimes not. Nuclear energy decisions often require complicated tradeoffs between such diverse things as economic benefits, technological development, environmental risks and possible long-term consequences for public health. These tradeoffs are basically value judgements. Different groups involved may disagree over the relative weight of the values and utilities involved. Eliciting, measuring and integrating these values in a decision can be made easier with decision analytic tools. Some of these will be briefly discussed.

The complexity of nuclear policy decisions is further enhanced by differences of opinion and judgement between the various stakeholders, i.e., national and local governmental authorities, industry and the public. Some decision analytic tools primarily focus on helping to improve the communication between the involved parties. One example of this category will be discussed.

When describing these techniques I will occasionally refer to the extensive literature on human judgement. Most of the tools mentioned above assume that people are reasonably adequate judges of probabilities and utilities and that, ideally, they combine these elements according to normative models. On the one hand, the literature on cognitive errors and biases underlines the necessity of using decision analytic tools to help improve the quality of human decision-making. On the other hand, the notion that systematic errors and inferential biases characterize human thinking could also imply that the tasks required by decision analysis (e.g., assessing probabilities, utilities and values) simply cannot be adequately performed by people. The latter issue will be discussed later in this chapter.

This chapter will emphasize the role of the expert. Unlike much of the previous chapters I will focus not on lay attitudes and judgements but on decision-making by experts. One theme, however, will return. Some of the decision analysis tools primarily aim to improve communication between the various stakeholders. (The latter issue was discussed in Chapter 8, which focused on risk communication.)

In this chapter I will first discuss applications of decision analysis in the area of risks and probabilities: *scenario development* and *risk analysis*. Next I will turn to methods that focus on value and/or utility; both *cost–benefit analysis* and *social and environmental impact assessment* will be briefly discussed. Finally, I will deal with some tools that aim to aid the communication between the involved parties and hence the decision-making process; one of these value-oriented decision tools will be presented. The actual and possible roles of the various techniques will be briefly touched upon.

Scenarios as decision-making tools

An important step in the evaluation and development of policy measures is the construction of scenarios describing possible futures under various conditions. The development of scenarios has typically been applied to long-term effects of energy policy alternatives. Scenarios describe the effects of alternative policy options for periods of ten to fifty years. Moreover, given the complexity of most energy policy issues, most scenarios are based on both empirically tested relationships between variables and more tentative 'educated guesses'. The long time-horizon accentuates the effects of this complexity. Scenarios always involve a variety of expectations and assumptions about variables such as population growth, changes in economic structure, technological risks, economic growth, changes in life-style, and the availability of non-renewable energy sources.

Scenarios at first focused on the interaction of environmental and economic development and consisted of simple input-output matrices. Later developments more explicitly aimed to provide an integrated analysis of economic and environmental processes. The SEAS model (strategic environmental assessment system) is an early example of models attempting to predict possible long-term future developments. The model includes a wide range of environmental variables and their interactions (see, e.g., Ratick and Lakshmanan, 1980). The model was developed as a technique to analyse, evaluate and also develop long-term policy alternatives. The ambitious scope of the model led to the inclusion of a wide array of variables, equations and data sets. Not surprisingly, this huge complexity, the relative inaccessibility and the substantial maintenance requirements (updating, adaptation to changing policy requirements etc.) resulted in a model with limited applicability.

A more recent generation of models that deal with scenarios is less ambitious. They usually focus on specific policy issues. Scenarios for long-term energy policy (e.g., Sassin et al., 1983) and acidification in Europe (Alcamo et al., 1985) are examples of this category. Another area of application is decisions about the storage of high-level radio-active waste. These are likely to rely more and more on scenario studies. The latter attempt to describe the possible health and environmental risks of storage facilities and assess the probability of 'worst case' scenarios. These scenarios need to use extremely long time-horizons. Not surprisingly, even these less ambitious scenarios are affected by the uncertainties and complexities of predicting possible futures.

Scenario studies usually provide an overview of possible (policy) measures and their consequences. They generally aim to provide a context in which alternative measures can be compared. The aim of most scenarios is simply to help structure the problem and to assist in the selection of alternatives to reach specific goals. The complexity of energy-policy issues implies that scenario studies in this domain tend to simplified models of possible future developments and can best be described as helping to focus the attention on crucial variables and on the important steps in the decision process. Some authors have attempted to develop taxonomies of scenario studies (e.g., Ducot and Lubben, 1980). Generally, however, this field of study lacks standard methodologies (see also Biel and Montgomery, 1986). Most research fails to provide a detailed account of the methodological aspects of the scenario study. Moreover, the variety of techniques being used is quite substantial and most of these are not accompanied by theoretical justification and/or empirical validation.

One attempt to use scenario construction as a tool in decision-making concerns the storage of high-level nuclear waste. PAGIS (performance assessment of geological isolation systems) is the first attempt at a multinational safety assessment of high-level radioactive waste. This effort, financed by the Commission of the European Communities, evaluates the possible radiological impact on humans of disposal in caly, granite and salt. For each of these disposal options the research team selected a reference case, computed the radioactivity decay and total decay heat for up to one million years (see Peters, 1989). Detailed models of the barrier system and the steel canisters holding the waste were used to produce predictions for, for example, expected dose rate for several scenarios. The scenarios also include possible developments such as a fault crossing the repository, the consequences of climatic changes, human intrusion and brine intrusion in salt mine. PAGIS is principally a pre-regulatory exercise and an example of a less elaborate, more specific type of scenario study as compared with the other energy policy scenarios mentioned.

The lack of generally accepted methodologies and the need to reduce complicated issues to manageable proportions imply that the study and development of scenarios leave considerable space for intuition. Not surprisingly, some researchers in the field of psychological decision theory argue that the incorporation of cognitive factors that can influence the quality of scenario construction is essential (Jungermann, 1985; Vlek and Otten, 1987). Most research on scenarios, however, is conducted by economists, engineers and planners. Contributions from the social and behavioural sciences are relatively rare. Psychologists usually refer to the outcomes of research on cognitive heuristics and biases as a possibility for improving the quality of scenario construction. Psychological research on heuristics and biases tends to be dominated by attempts to expose systematic errors and inferential biases in human reasoning. Below, I will briefly mention a number of factors that constitute potential difficulties in scenario construction.

Probability assessment Information about the probability of the various outcomes of a scenario may be available prior to the scenario development; it may be gathered during the construction of the scenarios; or it may be inferred from similar cases. In all these cases it is important to know how people use information about the probability of outcomes (e.g., the relative frequency of the outcomes in the past) or how people estimate the probabilities of specific future outcomes.

There are various factors that can influence the accuracy of people's estimation of probability. Most of these factors are related to the incorrect use of statistical evidence. Prior probabilities, the effect of sample size and conjunctive probabilities seem not to be clearly understood and are used incorrectly or are underutilized in probability estimates. Generally, intuitive judgements of probability seem to be strongly influenced by relatively concrete and vivid instances. The availability heuristic is used when estimates of frequency or probability are affected by the relative ease with which instances can be recalled from memory, or by the salience of examples in one's immediate environment. This cognitive salience makes it easier to retrieve instances from memory, and will influence judgements of the frequency of specific events. For example, seeing a television programme on nuclear hazards is likely to increase the perception of nuclear risks, at least temporarily. Such experience mainly affects lay people; people involved in the nuclear industry are likely to be aware of accident-free facilities and thus will be influenced more by this specific experience. In their case judgement is likely to be influenced by the inability to imagine things that have not happened before. In other words, in saying 'it won't happen to me' people show a tendency to view themselves (or their facility) as relatively immune to hazard. The most frustrating effect of lack of imaginability is that it could lead to a failure to appreciate the limits of available data and, as a consequence, may even lull people into a false sense of complacency (Slovic, Fischhoff and Lichtenstein, 1982). A further factor that could affect probability estimates concerns the excessive weight given to instances that *conform* to prior expectations. All in all, it is clear that people find it difficult to make probability estimates. This is also illustrated by the ease with which concrete instances can affect probability estimates and by the substantial effects caused by the introduction of anchor values (e.g., Kahneman, Slovic, and Tversky, 1982; van Schie and van der Pligt, 1992).

Some cautionary notes are in order, however. Not enough is known about the general applicability of the findings obtained in this research area. Questions such as whether the stimuli and designs used in experimental demonstrations are typical or exceptional need further attention. Another issue that needs to be resolved is whether people can be trained to avoid the various errors and biases in their use of statistical information and their estimates of probability (see, e.g., Jepson, Krantz and Nisbett, 1983).

Scope of scenarios A suggestion frequently made by social and behavioural scientists is that social and psychological impacts should be included in scenario studies. Attempts to include these aspects have been made in the areas of 'Environmental Impact Assessment' and 'Technology Assessment'. Inclusion of these aspects in scenario studies would considerably increase the complexity and uncertainty of scenarios. For this reason it seems advisable to limit social impacts to relatively obvious and major social trends. The fact that most scenarios attempt to deal with long-term futures restricts the number of social and psychological impacts that can be included in such scenarios. Most long-term social and political implications are extremely difficult to incorporate in scenarios. Generally, people acknowledge the need to incorporate specific social aspects (see, e.g., Jungermann, 1985), but nothing much is done. This seems to be mainly because of methodological problems (e.g., defining and measuring the relevant concepts) and the difficulties in predicting social and political changes and their impacts. Decisions about the consequences which will be incorporated in scenario studies are also related to values. In the second half of this chapter I will discuss value related issues in more detail.

Scenario studies usually focus on possible macro-economic and social consequences and are sometimes used to compare the consequences of possible energy futures. The extent to which the outcomes of scenario studies are used in policy decision-making tends to vary enormously, often for the wrong reasons. When scenario outcomes are in accordance with intended policies they function primarily as 'scientific' support for these policies and receive considerable attention. In these circumstances policy-makers tend to interpret the scenarios as objective, extremely reliable forecasts. They dislike any references to the uncertainty of the predicted outcomes. In other cases, outcome of scenario studies which would imply considerable policy changes have been ignored (see van der Pligt, 1989).

Because of the many uncertainties associated with scenario construction, the method has been severely criticized. On the other hand it can be argued that attempts to consider uncertain futures in a systematic and analytic way, even with incomplete and uncertain data, are still an improvement compared with holistic judgements about possible futures.

Risk assessment and policy decision-making

Risk assessment techniques are usually applied to more specific aspects of nuclear energy, such as possible accidents and their consequences. The scope of risk assessment techniques is thus much more limited than that of scenario construction. As was argued in Chapter 2, the risks associated with several recently introduced technologies led to the development of quantitative risk assessment techniques in order to base policy on 'objective' facts. The expectation that objectivity could be achieved has since been shown to be rather optimistic. Risk assessments generally leave many degrees of freedom to the analyst and the results of an assessment can easily be influenced by assumptions. The large uncertainties involved necessarily push the evidence out of the realm of 'facts' and into the realm of 'opinion'. One reason for this uncertainty is the limited knowledge of dose–response relations that should underlie risk assessments. As Chapter 2 showed, considerable research has been directed at risks associated with nuclear power generation but even in this area knowledge about a number of aspects is relatively limited (e.g., the long-term effects of routine low-level emissions). In this section I will discuss the impact of risk assessment techniques on policy decision-making.

Not surprisingly, relatively little research has been done on methodological and procedural aspects of risk assessments and their relationship to policy decisions. Lathrop and Linnerooth (1983) studied the siting of a liquified natural gas facility on the California coast and concluded that the political decision-making process bears little resemblance to the analyst's perspective. They suggest procedural changes (e.g., rules of evidence or standards to which risk assessments must adhere) in order to improve the use and effectiveness of risk assessments in the development of public policy.

Quantified risk assessment is an indispensable element in predicting the likelihood of the many possible hazards that can arise in complex facilities such as nuclear power plants and waste repositories. Generally, the technique should be used with caution. In deciding whether a specific risk is tolerable or not quantified risk assessment should be only one of the inputs into what is essentially a political decision. Some years ago, Lathrop and Linnerooth (1983) noticed a substantial gap between our knowledge about risk assessment and the ways in which these risks are handled by public institutions in their policy decision-making. They argued that the political processing of uncertainties, where risks and their acceptability are negotiated sequentially by public officials, industrial representatives and public interest groups simply does not allow the people involved to operate with an open recognition of the scientific uncertainties underlying their policy preferences and/or decisions.

Some attempts have been made to assess the strengths and limitations of risk assessment and its relevance to the wider decision-making process about the nuclear risks faced by workers and the public (HMSO, 1989). This application concerned the prediction of overall risk from a nuclear plant or equipment; decisions about whether a specific risk is acceptable or tolerable; and the extent to which the ranking of risks is useful for general policy decision-making.

This exercise suggested that risk assessments are of limited value in guiding policy decisions because of the many qualitative factors involved, which make it difficult, impossible or even misleading to relate one specific risk to another. Unquantifiable factors include such issues as the low public acceptability of nuclear risks. Other difficult aspects include the assessment of societal risk (which takes account of the total harm that might be suffered by whole communities). There seem to be too many factors involved in assessing all of these risks and most of them involve qualitative judgements. As a consequence, integration of risk assessment techniques in policy decision-making procedures is likely to be complicated.

Notwithstanding these difficulties, environmental policy is likely to depend more and more on quantitative risk assessments, especially in the areas of nuclear energy and other hazardous industrial activities. Fortunately, there seems to be a general tendency to make the practice of risk assessment more consistent throughout the various policy areas (see also Russell and Gruber, 1987). Applications, however, are often hampered by the necessity to include the above-mentioned qualitative judgements. The latter always include values and ethical considerations. In the next section I will discuss some techniques that emphasize the values or utilities of decision alternatives. Of course these techniques also involve probabilities as was discussed in the previous section. Two techniques are of particular interest to us. First, cost–benefit analysis; second, social and environmental impact assessment. Both techniques are related to multi-attribute theory.

Multi-attribute utility theory

When evaluating energy policy alternatives or sites for nuclear facilities, decision-makers must make complex tradeoffs among costs, risks and benefits. In these instances an overall evaluation is extremely complex, especially if the various value dimensions are in conflict. To help decision-making in such complex tasks, multi-attribute utility theory (MAUT) was developed. In MAUT the task is broken down into separate attributes and these are evaluated. Next, tradeoffs among attributes are operationalized as weights assigned to each of the attributes. Finally, formal models are used to reaggregate the evaluations on the various attributes. Most MAUT procedures include the following five steps:

(1) define alternatives and value-relevant attributes
(2) evaluate each alternative separately on each attribute
(3) assign relative weights to the attributes
(4) combine the weights of attributes with the evaluation of the attributes to obtain an overall evaluation of each of the alternatives
(5) perform sensitivity analysis and make recommendations (see von Winterfeldt and Edwards, 1986, p. 272).

Sensitivity analysis consists of changing the numbers and structures in the procedure with the aim of gaining insight into the nature of the problem and to help find a structure that constitutes an adequate representation of the problem.

A variety of methods are available to construct value or utility functions, weighting the various attributes and aggregating the values or utilities across attributes. These methods have been mainly applied to relatively complex problems. The idea of attempting to decompose complex decision problems into separate elements to be combined later to obtain an overall evaluation is essential to both techniques to be discussed in the next two sections. Hence the inclusion of MAUT-techniques in this chapter.

There are several ways in which values might be inadequately considered in decision-making. Three will be briefly discussed in this section: (1) relevant values or utilities may be overlooked; (2) there may be uncertainty about the relevance of various values; (3) there may be difficulties in assessing values or utilities.

The quality of decisions can be seriously affected if a decision-maker ignores important positive or negative features of one or more of the choice options. Some researchers (e.g., Janis and Mann, 1977) argue that the quality of decision-making should primarily be judged by the extent to which the processes are complete and unbiased. One potential bias is introduced in the tendency to look for confirming evidence; i.e., negative features of favoured options tend to be overlooked. Apart from the failure to include value considerations, it occasionally happens that people are unsure about what values are relevant. Fischhoff, Slovic and Lichtenstein (1980) point out that this is especially the case with unfamiliar and complex issues such as those related to scenario construction or complicated decisions with multiple objectives. Because of lack of experience and the complexity of the many possible consequences, people might not even know how to begin thinking about certain issues. Many of today's issues fall into this category, such as nuclear energy and nuclear waste with their possible long-term consequences for the environment and public health.

Even if it is known which values are relevant to a specific decision it might be difficult to assess these values. In some decision situations one will have to estimate value outcomes indirectly. Sometimes, however, one may rely upon feature(s) that have little diagnostic significance. Procedures to elicit underlying value or utility structures in order to make these more explicit seem especially important in the context of scenario construction. Quite often, disagreement about the validity of scenarios is related to differences in values, partly because values determine to a large extent the focus of scenario studies. Next I will discuss two techniques related to multi-attribute utility theory: cost–benefit analysis and impact assessment.

Cost–benefit analysis

One of the decision analysis techniques frequently employed to improve the quality of decision-making is cost–benefit analysis. The technique of cost–benefit analysis can be broken down into six steps:

(1) enumerate all the adverse consequences of a particular course of action
(2) estimate the probability that each of these consequences will occur
(3) estimate the costs should a particular consequence occur
(4) calculate the expected loss from each consequence by multiplying the costs by its probability of occurring
(5) compute the expected costs or losses for the course of action by summing the losses associated with the various consequences
(6) repeat the procedure for benefits.

Many different techniques go by the name cost–benefit analysis. The label has been used for almost any explicit consideration of the (monetary) advantages and disadvantages of one or more options. Generally, these values of the good and bad consequences of each option are assessed with the tools of economic theory (e.g., market behaviour). Cost–benefit analysis, however, involves several strong assumptions. It assumes, for example, that all important consequences of an action can be enumerated in advance, that the probability of their occurrence can be reliably estimated, and that different kinds of costs can be compared (for instance, in the context of rescue operations: financial versus pain, suffering and loss of life). All of these assumptions can and have been questioned (see also Chapter 2).

For a long time it seemed that alternative energy sources and energy conservation schemes would not result in net benefits. During the 1970s consensus among experts indicated that the costs of nuclear power would at least be equalled by the extra national income which could be brought about by its use. This changed during the 1980s because of the unexpectedly high costs of dismantling nuclear power stations and the falling prices of non-renewable energy sources. The 1990s, however, could provide a different picture, mainly because of the greenhouse effect caused by fossil fuels.

Apart from these changes, potential income as measured by the Gross National Product (GNP) has some peculiar features. Let us consider the following example: if health risks, and therefore total expenditures on prevention and care, are related to income growth, then these expenditures will increase GNP even though welfare (in terms of adverse health effects) has been reduced.

More serious problems are associated with the costs of nuclear energy. These should include radiation hazards, the costs of the disposal of nuclear waste and the dismantling of nuclear power stations. Furthermore, issues such as the costs of accounting for down to perhaps the last 100 kilograms of plutonium, the risks of nuclear proliferation and the costs for future generations make cost–benefit analysis a highly complicated task. The essence of cost–benefit analysis is quite straightforward: it establishes all the costs and benefits relating to a given alternative. The alternative should be selected only if benefits exceed costs.

Cost–benefit analysis has a potential use in ordering the decision-making process and making explicit the various attributes or aspects that are of relevance. Many decisions are taken without even these considerations. It seems therefore that cost–benefit analysis could have a valuable function. The problems for this method are two-fold. First, it is difficult to apply in contexts where the costs and benefits are not well-defined. This could be the result of the necessity to include a wide variety of potential costs, e.g., health risks, nuclear proliferation and costs to future generations. All of these potential costs are linked to conflicts about probabilities.

Secondly, the technique is confronted with the need to compare disparate costs. Generally, cost–benefit analysis aims to include all consequences amenable to economic valuation and to exclude all others (see, e.g., Parish, 1976). Many practitioners tend to evaluate only those commodities and services that have readily measurable market values (for example construction costs, wages). Indirect economic evaluation with the help of demand principles, shadow prices, and the like are sometimes used to extend the range of consequences to which a monetary value may be attached. There is disagreement about how far cost–benefit analysis should be extended into the realm of social and political consequences (see, e.g., Mishan, 1976; Parish, 1976). There is, however, increasing pressure to include these 'soft' values.

One of the consequences of this development is that behavioural scientists have become more involved in environmental cost–benefit analysis. This increased involvement was partly because of the need to develop benefit measures and non-market pricing techniques for valuing environmental attributes. For instance, economic assessments of the impact of noise at first tended to focus on aspects such as property values; i.e., the relationship between traffic noise and residential property values as a possible basis for cost–benefit analyses of noise-abatement projects. Next, epidemiological studies attempted to extend the scope of cost–benefit analysis by quantifying the consequences of noise on public health. This attempt had mixed success because of the complexity of the issue and the many confounding possible antecedents. Finally, psychologists entered this field with the aim of developing non-market valuation techniques to assess environmental quality. Difficulties in the quantification of costs such as stress-related psychological and physical reactions to reduced environmental quality led to a search for techniques that could provide quantitative, monetary estimates of environmental quality. From the mid 1980s onwards economic and psychological research in this area has tended to emphasize the measurement of benefits of environmental policy. This emphasis is the result of the need of policy-makers to base their policy measures on extensive cost–benefit analysis.

Impact assessment

The previous section argued that there has been considerable pressure to widen the scope of cost–benefit analysis to also include less tangible consequences. This approach tends to be accompanied by considerable difficulties in the quantification of environmental, social, and psychological costs. It seems best used as a tool to help take into account all possible consequences of policy alternatives. The levels of uncertainty associated with some of the impacts are, however, substantial. Richardson, Sorensen and Soderstrom (1987) proposed to focus on three separate levels of social and psychological impacts: the individual, family and community level. Their study dealt with nuclear accidents, but their approach could also be applied to siting procedures. For each of the three levels they used specific indicators, ranging from health and well-being at the individual level, disruption of family life at the second level and perceived or expected community impacts at the third level.

Impact assessment techniques rely to a large extent on self-report measures. As Chapter 6 showed, self-reports have serious disadvantages, especially when respondents' replies could influence the outcome of a siting procedure or the compensation to be paid after an accident. The main benefit of these more qualitative approaches is that they give an indication of the perceived or expected consequences as seen by the affected people in the locality. However difficult to quantify, it seems necessary to be aware of these aspects and to take them into account in decision-making procedures. Impact assessment techniques could possibly improve the communication between experts, policy-makers and lay people by making the former groups more aware of the impacts as perceived by lay people. There are several techniques that aim to improve communication and help societal decision processes. One of these will be discussed in the next section.

Value-oriented social decision analysis

Chen and Mathes (1989) argue that many public policy decisions cannot be easily reduced to simple 'either-or' propositions. This certainly applies to energy policy decisions. According to Chen and Mathes this tendency to oversimplify can take a variety of forms. One is to simplify the issue in such a way that it becomes one of the interests of the community as a whole versus the interests of the neighbourhood. Another way is to turn the issue into an ideological issue which conforms to an accepted opposition of values in the community (such as pro-business or development versus quality of life).

These simplifying tendencies could lead to less appropriate solutions. In order to help improve the quality of policy decision-making Chen and Mathes attempted to develop a decision-making aid which helps to define a specific problem and the related issues in a way that reflects both the complexity of the issue and the various values of the interest groups involved. Their technique aims to recognize that various persons and groups have different understandings of the problem, different interpretations of the possible solutions, and different values for evaluating those solutions. The method they developed aims to improve the effectiveness of decision-making on public policy issues by improving communication between the various stakeholder groups. Moreover, the method is expected to be most efficient in situations where the nature of the problem is complex and it is difficult to decide which issues are involved. The method, called Value-Oriented Social Decision Analysis or VOSDA, is based upon multiple criteria decision-making techniques described earlier in this chapter. In other words, the technique helps to identify alternative options and their consequences, judge the likelihood of these outcomes and establish the desirability of these possible outcomes. The method expands this basic technique by including the relevant stakeholder groups and enhances discussion and communication between these groups. The decision analysis for each of the stakeholders would thus be based on multi-attribute utility theory, i.e., the overall utility (U) would be $\Sigma_i \Sigma_j W_i P_{ij} U_{ij}$, where:

U is the alternative's overall expected utility
W_i is the weight assigned to attribute i
P_{ij} is the probability of event j in terms of attribute i
U_{ij} is the utility of event j in terms of attribute i

The overall expected utility for a specific alternative consists of the sum of the expected utilities of that option for each attribute, adjusted according to the weights assigned to the attributes. The decision aiding technique is to complement the traditional political decision-making process. The output of a case consists of descriptive (and graphic) representations of the various positions on the policy alternatives as determined by the decision analysis. This diversity of opinion is then fed into the normal political policy-making process.

Applications of this procedure look promising but are also extremely time consuming. VOSDA could, however, play an important role in enhancing mutual understanding between the various parties or groups involved. In this way a combination of decision analysis techniques and decision conferencing (Philips, 1984) could help to improve the quality of policy decisions and help to depolarize debates about important issues such as the future production of electric power.

Conclusions

The 1980s saw a growing number of applications of notions developed in the area of decision theory to help formulate policy decisions. In this chapter we focused on nuclear energy policy and briefly reviewed three techniques: scenario construction, risk assessment and some cost–benefit analysis techniques. The application of decision-analytic concepts and tools is a difficult and complex task. Political factors seem as important as the quality of the scientific basis of the research in determining the impact of research findings on policy development. Several techniques have helped to formulate policy measures (e.g., scenario construction, risk assessments).

Although the methods described in this chapter have their shortcomings, mainly because of the complexity of the issues at hand, each of them could be valuable in the context of energy policy decision-making. First, they could help us to look at possible long-term consequences in a more systematic way. Second, multi-attribute approaches could help us to clarify decision problems and identify the relevant consequences of alternatives. As was argued before, most of these techniques are relatively new and applications to extremely complicated issues such as those discussed in this book are still relatively limited.

A third possible benefit of the approaches described in this chapter is improved communication between the various stakeholders. The various multi-attribute techniques could help to make explicit the differences between the interest groups. By decomposing complex issues these techniques could help to identify differences in perceptions (of the facts and values) of the issue. These techniques could also help to find alternative solutions, and help in the selection and weighting of criteria used to evaluate the alternatives. Improved communication could enhance the rationality and efficiency of the decision-making process. The often polarized nuclear debate could well do with techniques that increase the likelihood of mutual understanding and finding mutually acceptable solutions.

10

Conclusions

During the 1970s and 1980s nuclear energy generated a major and sustained controversy in most countries. Strong public opposition to nuclear energy led to diverging views between, on the one hand, local politicians and, on the other, central governments and energy authorities. Billions of dollars were lost. In such difficult circumstances policy-makers often turn to psychologists to 'solve' problematic situations. This is exactly what happened in the nuclear energy issue. Initially, psychologists became involved to explain to the public that low probabilities were nothing to worry about. Later, psychological research focused more on the how and why of public acceptance and on policy decision-making procedures to help reach better and more acceptable solutions.

It is more than likely that the standstill of most nuclear energy programmes since the late 1980s will come to an end. Environmental changes such as global warming are likely to lead policy-makers to reconsider the nuclear option. What has been learned in the past decades? Could psychology play a role in future decision-making and help to prevent a renewed polarization of the nuclear debate? This book has attempted to describe psychological research on the nuclear issue. The major conclusions of this research should provide a basis for future policy decision-making.

One of the major contributions of psychological research on the nuclear issue is that it points to important differences in the frame of reference of experts and lay people. Specific characteristics of the risks associated with nuclear energy (for example the catastrophic nature of the consequences, limited experience with the consequence, the perceived limited control over these consequences by both the relevant authorities, science and experts) play a major role in the public's acceptability of these risks. To a certain extent these worries have proved to be right. Major accidents in the USA and the Soviet Union have underlined public uncertainty about nuclear technology and have led to major shifts in public opinion, especially in countries near to the accident sites.

Psychological research has also revealed that local attitudes towards nuclear power stations and nuclear waste facilities are perfectly reasonable from a local perspective. In other words, explaining local resistance in terms of irrational fears is unlikely to help solve the issue. A more careful analysis has shown considerable discrepancy between (local) costs and (national) benefits for those locally affected by nuclear policy decisions. Equity is thus a major issue in public acceptance of nuclear facilities. Several approaches could help to restore the balance between costs and benefits. All of these have important consequences for policy decision-making. Respect for the public's perspective seems essential and a prerequisite for successful policy development. Increased participation of the public in nuclear decision-making could be the only way to repair a rather unfortunate history of communication with the public, which resulted in public distrust of the relevant regulatory authorities.

For many years the communication of nuclear risks has been relatively poor. Risk communication efforts providing *general* information about nuclear energy risks have been limited. Risk communication about these complex, technical issues has proved to be extremely difficult. Major lessons learned include the need to present information that goes beyond mere numerical estimates of the risks. Qualitative aspects of the risks involved are the prime determinant of public reactions to nuclear energy. Moreover, research also indicates that one should be extremely cautious in using risk comparisons (e.g., comparing the risk of a nuclear accident with the risk of being struck by lightning). These comparisons tend to be seen as misleading and, as a consequence, are likely to increase public opposition. A careful consideration of the dimensions or qualitative characteristics of risk that determine public acceptance seems a more promising approach.

Unfortunately, risk communication in emergency situations after large scale nuclear disasters also has a poor track record. In this book I presented an overview of risk communication efforts after the TMI and Chernobyl accidents. Both accidents resulted in inadequate efforts to communicate the risks to the public. The major shortcoming was substantial inconsistencies over time because of the large number of parties involved and/or lack of knowledge. Again, these episodes did not improve the credibility of the relevant authorities. Psychology could help to improve future risk communication efforts. Communication aims to inform but it should also bridge the gap between different perspectives. Some of the results described in this book should help to bridge the gap between the experts, policy-makers and lay people.

Psychology could also help to improve the quality of policy decision-making. In the last chapter I briefly presented some new techniques that could enhance the completeness and rationality of policy decision-making. These tools could help decision-makers to look at possible long-term consequences in a more systematic way. These tools could also help to decompose the decision problems. This should help the decision-making process to focus on the relevant aspects of technological developments. Finally, the methods described in the last chapter could help to elicit the relevant *values* underlying nuclear policy decisions. Policy decision-making controversies are not only based on different perceptions of facts but are also related to different value systems. Decision support should help to make these differences more explicit and improve the quality of policy decisions. Moreover, explicit consideration of the relevant facts and values is also likely to help (risk) communication about possible nuclear futures.

Above all, psychology could contribute to the depolarization of the nuclear debate, and help to find acceptable solutions to our future energy requirements.

Bibliography

Adler, A. (1943). Neuropsychiatric complications in victims of Boston's Cocoanut Grove disaster. *Journal of the American Medical Association*, 123, 1098–1111.
AIF (1981). *A Proposed Approach to the Establishment and Use of Quantitative Safety Goals in the Nuclear Regulatory Process*. Washington, DC: Committee on Reactor Licensing and Safety; Atomic Industrial Forum.
Alcamo, J., Hordijk, L., Kamari, J., Kauppi, P., Posch, M. and Runca, E. (1985). Integrated analysis of acidification in Europe. *Journal of Environmental Management*, 21, 47–61.
Allensbach, Institut für Demoskopie (1987). *Public Opinion Poll on Nuclear Energy after Chernobyl*. Report for the Atomic Industrial Forum, FRG: Allensbach.
Allport, G.W. (1935). Attitudes. In Murchison C. (ed.) *Handbook of Social Psychology*. Worcester, Mass: Clark University Press.
Appley, M. and Trumbull, R. (eds) (1967). *Psychological Stress*. New York: Appleton, Century, Crofts.
Atom (1990). Radioactive waste: good progress towards a solution. *Atom*, 404, June 1990, p. 6.
Atom (1991). US public expect a nuclear future. *Atom*, 413, May 1991, p. 3.
Averill, J.R. (1973). Personal control over aversive stimuli and its relationship to stress. *Psychological Bulletin*, 80, 286–303.
Bachrach, K.M. and Zautra, A.J. (1986). Assessing the impact of hazardous waste facilities: psychology, politics, and environmental impact statements. In Lebowitz, A.H., Baum, A. and Singer, J.E. (eds) *Advances in Environmental Psychology*, 6, 71–88.
Baum A., Fleming, R. and Davidson, L. (1983). Natural disaster and technological catastrophe. *Environment and Behavior*, 15, 330–54.
Baum, A., Fleming, R. and Singer, J.E. (1982). Stress at Three Mile Island: applying social impact analysis. *Applied Social Psychology Annual*, 3, 217–48. Beverly Hills, Calif: Sage.

Baum, A., Gatchel, R.J. and Schaeffer, M.A. (1983). Emotional, behavioral, and physiological effects of chronic stress at Three Mile Island. *Journal of Consulting and Clinical Psychology*, 51, 565–72.

Baum, A., Grundberg, N. and Singer, J.E. (1982). The use of psychological and neuroendocrinological measurement in the study of stress. *Health Psychology*, 1, 217–36.

Baum, A., Singer, J.E. and Baum, C. (1982). Stress and the environment. In Evans, G.W. (ed.) *Environmental Stress*, 15–44. New York: Cambridge University Press.

Baumann, D. (1983). Determination of the cost effectiveness of flood hazard information programs. *Papers and proceedings of the Applied Geography Conference*, 6, p. 292.

Bertell, R., Jacobson, N. and Stogre, M. (1984). Environmental influence on survival of lowbirth weight infants in Wisconsin, 1963–1975. *International Perspectives in Public Health*, 1(2), 12–24.

Biel, A. and Montgomery, H. (1986). Scenarios in energy planning. In Brehmer, B., Jungermann, H., Lourens, P. and Sevon, G. (eds) *New Directions in Research on Decision Making*. Amsterdam: North-Holland.

Bishop, W.P. (1978). Observations and impressions on the nature of radioactive waste management problems. In Bishop, W.P., Hilberry, N., Hoos, I.R., Metlay, D.S. and Watson, R.A. (eds) *Essays on Issues Relevant to the Regulation of Radioactive Waste Management*. NUREG-0412 US Nuclear Regulatory Commission, Washington, DC.

Black, J.S. (1987). *Regulation and Public Involvement: the Case of Hazardous Wastes*. Paper presented at the American Association for the Advancement of Science, 14–18 February 1987, Chicago, Illinois (7 pp.).

Bord, R.J. (1987). *Public Cooperation as a Social Problem: the case of risky wastes*. Paper presented at the American Association for the Advancement of Science, 14–18 February 1987, Chicago, Illinois (9 pp.).

Bord, R.J., Ponzurick, P.J. and Witzig, W.F. (1985). Community response to low-level radioactive waste: a case study of an attempt to establish a waste reduction and incineration facility. *IEEE Transactions on Nuclear Science*, 32, 4466–71.

Bowonder, B. (1985). *Low Probability Event: a case study in risk assessment*. Paper presented at the workshop 'Risk Analysis in Developing Countries', Hyderabad, India, October.

Broadbent, D. (1971). *Decision and Stress*. New York: Academic Press.

Bromet, E. (1980). *Three Mile Island: mental health findings*. Pittsburgh, PA: Western Psychiatric Institute and Clinic.

Brooks, H. (1976). *Waste Management*. Paper presented at the International Symposium on the Management of Wastes from the LWR Fuel Cycle, Denver, Colorado, 12 July.

Brown, J., Henderson, J. and Fielding, J. (1983). *Differing Perspectives on Nuclear Related Risks: an analysis of social psychological factors in the perception of nuclear power*. Paper presented at the meeting of the Operational Research Society, September, University of Warwick, UK.

Burns, M.E. (1984). Striking a reasonable balance. In Harthill, M. (ed.) *HazardousWaste management: in whose backyard?* 185–200. (American Association for the Advancement of Science, selected symposium 88). Boulder, Colo: Westview Press.

Campbell, J. (1983). Ambient stressors. *Environment and Behavior*, 15, 355–80.

Carle, R. (1981). Why France went nuclear. *Public Power*, July/August, 58–85.

Carnes, R., Copenhaver, E., Reed, J., Soderstrom, J., Sorensen, J., Bjornstad, D. and Peelle, E. (1982). *Incentives and the Siting of Radioactive Waste Facilities (ORNL-5880)*. Oakridge National Laboratories, Oak Ridge.

Checkoway, B. (1981a). The politics of public hearings. *Journal of Applied Sciences*, 17, 566–82.

Checkoway, B. (1981b). Incentives and nuclear waste siting: prospects and contstraints. *Energy Systems and Policy*, 7, 323–51.

Chen, K. and Mathes, J.C. (1989). Value oriented social decision analysis: a communication tool for public decision making on technological projects. In Vlek, C. and Cvetkovich, G. (eds) *Social Decision Methodology for Technical Projects*, 111–32. Deventer: Kluwer Academic Publishers.

Cohen, B.L. and Lee, I.S. (1979). A catalog of risks. *Health Physics*, 36, 707–22.

Cohen, S. (1980). Aftereffects of stress on human performance and social behavior: a review of research and theory. *Psychological Bulletin*, 88, 82–108.

Cohen, S., Evans, G.W., Krantz, D.S. and Stokols, D. (1980). Psychological, motivational, and cognitive effects of aircraft noise on children: moving from the laboratory to the field. *American Psychologist*, 35, 231–43.

Cohen, S. and Spacapan, S. (1984). The social psychology of noise. In Jones, D.M. and Chapman, A.J. (eds) *Noise and Society* (221–45). New York: Academic Press.

Cohen, S. and Weinstein, N. (1982). Nonauditory effects of noise on behavior and health. In Evans, G.W. (ed.) *Environmental Stress* (45–74). New York: Cambridge University Press.

Collins, D.L., Baum, A. and Singer, J.E. (1983). Coping with chronic stress at Three Mile Island: psychological and biochemical evidence. *Health Psychology*, 2, 149–66.

Colson, J.P. (1977). *Le Nucleaire sans les Français*. Paris: Maspero.

Combs, B. and Slovic, P. (1979). Causes of death: biased newspaper coverage and biased judgments. *Journalism Ouarterly*, 56, 837–43.

Commission of the European Communities (1982). *Public Opinion in the European Community* (Report No. XVII/202/83-E). Brussels: Commission of the European Communities.

Covello, V.T. and Allen, F. (1988). *Seven Cardinal Rules of Risk Communication*. Washington, DC: US Environmental Protection Agency.

Covello, V.T., Sandman, P. and Slovic, P. (1988). *Risk Communication, Risk Statistics, and Risk Comparisons: a manual for plant managers*. Washington, DC: Chemical Manufacturers Association.

Covello, V.T., von Winterfeldt, D. and Slovic, P. (1986). Risk communication: a review of the literature. *Risk Abstracts*, 3, 171–82.

Cunningham, S. (1985). The public and nuclear power. American Psychological Association, *Monitor*, 16: 1, 17.

Cutter, S. and Barnes, K. (1982). Evacuation behavior and Three Mile Island. *Disasters*, 6, 116–24.

Daamen, D.D.L., Verplanken, B. and Midden, C.J.H. (1986). Accuracy and consistency of lay estimates of annual fatality rates. In Brehmer, B., Jungermann, H., Lourens, P. and Sevon, G. (eds) *New Directions in Research on Decision Making*, 231–43. Amsterdam: Elsevier Science Publishers.

Davidson, L.M., Baum, A. and Collins, D.L. (1982). Stress and control-related problems at Three Mile Island. *Journal of Applied Social Psychology*, 12, 349–59.

Davidson, L.M., Baum, A., Fleming, I. and Gisriel, M.M. (1986). Toxic exposure and chronic stress at Three Mile Island. In Lebovits, A.H., Baum, A. and Singer, J.E. (eds) *Advances in Environmental Psychology*, 6, *Exposure to Hazardous Substances: Psychological Parameters*, 35–46. Hillsdale, NJ: Erlbaum.

Dawes, R.M. and Smith, T.L. (1985). Attitude and opinion measurement. In Lindzey, G. and Aronson, E. (eds) *Handbook of Social Psychology*, 1, 509–66. New York: Random House.

Decima (1987). *A Study of Canadians' Attitudes toward the Use of Nuclear Energy to Generate Electricity in Canada*. Report #2383. Decima Research, Ontario.

Dew, M.A. and Bromet, E.J. (1985). A matter of perspective: community residents' and nuclear power plant workers' long-term reactions to the Three Mile Island accident. *Social Studies of Science*, October 1985, 24–7.

Dohrenwend, B.P., Dohrenwend, B.S., Kasl, S.V. and Warheit, G.J. (1979). *Report of the Task Group on Behavioral Effects to the President's Commission on the Accident at Three Mile Island*. Washington, DC: USA Government Printing Office.

Dohrenwend, B.P., Dohrenwend, B.S., Warheit, G.J., Bartlett, G.S., Goldsteen, R.L., Goldsteen, K. and Martin, J.C. (1981). Stress in the community: a report to the president's commission on the accident at Three Mile Island. In Moss, T.H. and Sills, D.L. *The Three Mile Island Nuclear Accident: lessons and implications*, 159–74. New York: Annals of the New York Academy of Sciences.

Douglas, M. and Wildavsky, A. (1982). *Risk and Culture*. University of California Press, Berkeley, Calif.

Drottz, B.M. and Sjöberg, L. (1990). Risk perception and worries after the Chernobyl accident. *Journal of Environmental Psychology*, 10, 135–50.

Ducot, C. and Lubben, H.J. (1980). A typology for scenarios. *Futures*, 12, 51–7.

Dupont, R.L. (1980). Nuclear phobia: phobic thinking about nuclear power. In

Nuclear Power in American Thought, 23–41, Washington, DC: Edison Electric Institute.

Dupont, R.L. (1981). The nuclear power phobia. *Business Week*, 7 September, 14–16.

Earle, T.C. and Cvetkovich, G. (1985). *Failure and Success in Public Risk Communication*. Paper presented at the Air Pollution Control Association Conference on Avoiding and Managing Environmental Damage from Major Industrial Accidents, Vancouver, BC, Canada, November.

Earle, T.C. and Cvetkovich, G. (1988). *Platitudes and Comparisons: a critique of current (wrong) directions in risk communication*. Bellingham, Washington: Western Institute for Social and Organizational Research.

Earle, T.C. and Lindell, M.K. (1982). *Public Perception of Industrial Risks: a free-response approach*. Battelle Human Affairs Research Centers, Seattle, Wash.

Edwards, W. (1954). The theory of decision making. *Psychological Bulletin*, 51, 380–417.

Eiser, J.R. and van der Pligt, J. (1979). Beliefs and values in the nuclear debate. *Journal of Applied Social Psychology*, 9, 524–36.

Eiser, J.R., van der Pligt, J. and Spears, R. (1988). Local oppositions to the construction of a nuclear power station: differential salience of impacts. *Journal of Applied Social Psychology*, 18, 654–63.

Emmings, A. (1989). The next step to radioactive waste management. *Atom*, 391, May 1989, 6–9.

Enbar, M. (1983). Equity in the social sciences. In Kasperson, R.E. (ed.) *Equity Issues in Radioactive Waste Management*, 69–93. Cambridge, Mass: Oelgeschlager, Gunn and Hain.

Ester, P., Mindell, C., van der Linden, J. and van der Pligt, J. (1983). The infuence of living near a nuclear power plant on beliefs about nuclear energy. *Zeitschrift für Umweltpolitik*, 6, 349–62.

Evans, G.W. (ed.) (1982). *Environmental Stress*. New York: Cambridge University Press.

Evans, G.W. and Cohen, S. (1987). Environmental Stress. In Stokols, P. and Altman, I. (eds) *Handbook of Environmental Psychology*, 1, 571–610. New York: Wiley.

Evans, G.W. and Jacobs, S.V. (1982). Air pollution and human behavior. In Evans, G.W. (ed.) *Environmental Stress* (105–32). New York: Cambridge University Press.

Eyre, B. and Flowers, R. (1991). Radwaste management in the UK – present status and future plans. *Atom*, 418, 6–10.

Fallows, S. (1981). The nuclear waste disposal controversy. In Nelkin, D. (ed.) *Controversy: politics of technical decision*, 87–110. Berverly Hills, Calif: Sage.

Festinger, L. (1957). *A theory of Cognitive Dissonance*. New York: Harper and Row.

Fischhoff, B. and MacGregor, D. (1983). Judged lethality: how much people

seem to know depends upon how they are asked. *Risk Analysis*, 3, 229–36.
Fischhoff, B., Slovic, P. and Lichtenstein, S. (1979). Which risks are acceptable? *Environment*, 21, 14–20.
Fischhoff, B., Slovic, P. and Lichtenstein, S. (1980). Knowing what you want: measuring labile values. In Wallsten, T. (ed.) *Cognitive Processes in Choice and Decision Behaviour*. Hillsdale, NJ: Erlbaum.
Fischhoff, B., Slovic, P., Lichtenstein, S., Read, S. and Combs, B. (1978). How safe is safe enough: a psychometric study of attitudes toward technological risks and benefits. *Policy Sciences*, 8, 127–52.
Fishbein, M. (1963). An investigation of the relationship between beliefs about an object and the attitude toward that object. *Human Relations*, 16, 233–40.
Fishbein, M. and Ajzen, I. (1972). Attitudes and opinions. *Annual Review of Psychology*, 23, 487–544.
Fishbein, M. and Ajzen, I. (1975). *Belief, Attitude, Intention and Behavior*. Reading, Mass: Addison-Wesley.
Fishbein, M. and Hunter, R. (1963). Summation versus balance in attitude organization and change. *Journal of Abnormal and Social Psychology*, 69, 505–10.
Fiske, S.T. and Taylor, S.E. (1984). *Social Cognition*. Reading, Mass: Addison-Wesley.
Fitchen, J., Heath, J. and Fessenden-Raden, J. (1987). Risk perception in community context: a case study. In Johnson, B.B. and Covello, V.T. (eds) *The Social and Cultural Construction of Risk: essays in the perception and selection of risks*. Dordrecht, Netherlands: Reidel.
Flavin, C. (1987). Chernobyl: the political fallout in Western Europe. *Forum for Applied Research and Public Policy*, Summer 1987, 16–28.
Fleming, R., Baum, A., Gisriel, M.M. and Gatchel, R.J. (1982). Mediating influences of social support on stress at Three Mile Island. *Journal of Human Stress*, September 1982, 14–22.
Fleming, R., Baum, A. and Singer, J.E. (1984). Toward an integrative approach to the study of stress. *Journal of Personality and Social Psychology*, 46, 839–52.
Flynn, C.B. (1979). *Three Mile Island Telephone Survey: a preliminary report*. Washington DC: US Nuclear Regulatory Commission.
Flynn, C.B. (1981). Local public opinion. In Moss, T.H. and Sills, D.L. (eds). The Three Mile Island nuclear accident: lessons and implications. *Annals of the New York Academy of Sciences*, 365, 146–58. New York: New York Academy of Sciences.
Flynn, C.B. (1982). Reactions of local residents to the accident at Three Mile Island. In Sills, D.L., Wolf, C.P. and Shelanski, V.B. (eds) *Accident at Three Mile Island: the human dimension*, 49–64. Boulder, Colo: Westview Press.
Foa, F.B. and Foa, U.G. (1980). Resource theory: interpersonal behavior as exchange. In Gergen, K.J., Greenberg, M.S. and Harthill, M. (eds) *Social Exchange: advances in theory and research*, 77–94. New York: Plenum.

Folkman, S. and Lazarus, R.S. (1980). An analysis of coping in a middle-aged community sample. *Journal of Health and Social Behavior,* 21, 219–39.

Fowlkes, M. and Miller, P. (1987). Community and risks at Love Canal. In Johnson, B.B. and Covello, V.T. (eds) *The Social and Cultural Construction of Risk: essays in the perception and selection of risks,* 55–78. Dordrecht, Netherlands: Reidel.

Freedman, J.L. and Sears, D. (1965). Selective Exposure. In Berkowitz, L. (ed.) *Advances in Experimental Social Psychology,* 2. New York: Academic Press.

Freudenburg, W.R. and Baxter, R.K. (1984). Nuclear reactions: public attitudes and policies toward nuclear power. *Policy Studies Review,* 5, 96–110.

Gardner, M.J. (1990). Results of a case-control study of leukaemia and lymphoma among young people near Sellafield nuclear plant in West Cumbria. *British Medical Journal,* 300, 423–29.

Glass, D. and Singer, J. (1972). *Urban stress.* New York: Academic Press.

Gleser, G., Green, B.L. and Winget, C.W. (1978). Quantifying interview data on psychic impairment of disaster survivors. *Journal of Nervous and Mental Disease,* 166, 209–16.

Gleser, G.C., Green, B.L. and Winget, C.W. (1981). *Prolonged Psychosocial Effects of Disaster: a study of Buffalo Creek.* New York: Academic Press.

Goldhaber, M.K., Houts, P.S. and Disabella, R. (1983). Moving after the crisis: a prospective study of Three Mile Island area population mobility. *Environment and Behavior,* 15, 93–120.

Granberg, D. and Halmberg, S. (1986). Preference, expectations, and voting in Sweden's referendum on nuclear power. *Social Science Ouarterly,* 66, 379–92.

Groth, A.J. and Schutz, H.G. (1976). *Voter Attitudes on the 1976 Nuclear Initiative in California.* Institute of Governmental Affairs, University of California, Davis.

Hance, B.J., Chess, C. and Sandman, P. (1987). *Improving Dialogue with Communities: a risk communicating manual of government.* Trenton, NJ: State Department of Environmental Protection.

Hare, F.K. and Aikin, A.M. (1984). Nuclear waste disposal: technology and environmental hazards. In Pasqualetti, M.J. and Pijawka, K.D. (eds) *Nuclear Power: assessing and managing hazardous technology.* 311–48. Boulder, Colo.: Westview Press.

Harnden, D.G. (1989). Leukaemia Clusters, paper presented at 'The effects of small doses of radiation', London, 1989. Reported in *Nuclear News,* April 1989, 86.

Harris and Associates (1976). *A Second Survey of Public and Leadership Attitudes toward Nuclear Power Development in the United States.* New York: Louis Harris and Associates.

HMSO (1988). *Chernobyl: the governments' response, minutes of evidence,* 2. London: HM Stationery Office, July, 1988.

HMSO (1989). *Ouantified Risk Assessment: its input to decision making.* Paper by the Health and Safety Executive. London: HM Stationery Office.

Hodler, T.W. (1980). Residents' preparedness and response to the Kalamazoo tornado. *Disasters*, 2, 44–9.
Hohenemser, C. (1988). The accident at Chernobyl: health and environmental consequences and the implications for risk management. *Annual Review of Energy*, 13, 179–228.
Hohenemser, C., Deicher, M., Ernst, A., Hofsäss, J., Lidner, G. and Recknagel, E. (1986). Chernobyl: an early report. *Environment*, 28, issue 4, 6–13, 30–43.
Hohenemser, C., Kasperson, R. and Kates, R.W. (1977). The distrust of nuclear power. *Science*, 196, 25–39.
Hohenemser, C., Kates, R.W. and Slovic, P. (1983). The nature of technological hazard. *Science*, 220, 235–71.
Hohenemser, C. and Renn, O. (1988). Shifting public perceptions of nuclear risk: Chernobyl's other legacy. *Environment*, 30, issue 3, 4–11, 40–5.
Houts, P.S. and Goldhaber, M.K. (1981). Psychological and social effects on the population surrounding TMI after the nuclear accident on March 28, 1979. In Majumdar, S. (ed.) *Energy, Environment and the Economy*. Pennsylvania Academy of Sciences.
Houts, P.S., Miller, R., Tokuhata, G. and Ham, K. (1980). *Health-related Behavioral Impact of the Three Mile Island Incident: parts I and II*. Harrisburg: Pennsylvania Department of Health.
Hu, T., Slaysman, K., Ham, K. and Yoder, M. (1980). *Health-related Economic Costs of the Three Mile Island Accident*. Harrisburg: Pennsylvania Department of Health.
Hughey, J.B., Lounsbury, J.W., Sundstrom, E. and Mattingly, T.J. (1983). Changing expectations: a longitudinal study of community attitudes toward a nuclear power plant. *American Journal of Community Psychology*, 11, 655–72.
Hughey, J.B., Sundstrom, E. and Lounsbury, J.W. (1985). Attitudes toward nuclear power: a longitudinal analysis of expectancy-value models. *Basic and Applied Social Psychology*, 6, 75–91.
Interagency Review Group on Nuclear Waste Management (1979). *Report to the President* TID-29442. Springfield, Va: National Technical Information Service.
Janis, I.L. (1972). *Victims of groupthink*. Boston, Mass: Houghton Mifflin.
Janis, I.L. (1982). *Groupthink*, 2nd edition. Boston, Mass: Houghton Mifflin.
Janis, I.L. and Mann, L. (1977). *Decision Making*. New York: Free Press.
Jepson, C., Krantz, D.H. and Nisbett, R.E. (1983). Inductive reasoning: competence or skill? *Behavioral and Brain Sciences*, 6, 494–501.
Johnson, B.B. (1987a). Public concerns and the public role in siting nuclear and chemical waste facilities. *Environmental Management*, 11, 571–86.
Johnson, B.B. (1987b). Accounting for the social context of risk communication. *Science and Technological Studies*, 5, no. 3/4, 103–11.
Jungermann, H. (1985). Psychological aspects of scenarios. In Covello, V.T., Mumpower, J.L., Stallen, P.J.M. and Uppuluri, V.R.R. (eds) *Environmental*

Impact Assessment, Technology Assessment, and Risk Analysis, 219–36. Berlin: Springer Verlag.

Kahneman, D., Slovic, P. and Tversky, A. (1982). *Judgment under Uncertainty: heuristics and biases*. Cambridge: Cambridge University Press.

Kaplan, H.B. (1983). *Psychological Stress*. New York: Academic Press.

Kasperson, R.E. (1980). The dark side of the radioactive waste problem. In O'Riordon, T. and Turner, K. (eds) *Progress in Resource Management and Environmental Planning*, 2, 133–63. Chichester: Wiley.

Kasperson, R.E. (1985). *Rethinking the Siting of Hazardous Waste Facilities*. Paper presented at the conference on Transport, Storage, and Disposal of Hazardous Materials. International Institute for Applied Systems Analysis, Vienna, July 1985.

Kasperson, R.E. (1986). Six propositions on public participation and their relevance for risk communication. *Risk Analysis*, 6, 275–81.

Kasperson, R.E., Berk, G., Pijawka, D., Sharaf, A.B. and Wood, J. (1980). Public opposition to nuclear energy: retrospect and prospect. *Science, Technology and Human Values*, 5, 11–23.

Kasperson, R.E., Derr, P. and Kates, R.W. (1983). Confronting equity in radioactive waste management: modest proposals for a socially just and acceptable program. In Kasperson, R.E. and Berberian, M. (eds) *Equity Issues in Radioactive Waste Management*, 331–68. Cambridge, Mass: Oelgeschlager, Gunn and Hain.

Kates, R.W. and Braine, B. (1983). Locus, equity, and the West Valley nuclear wastes. In Kasperson, R.E. and Berberian, M. (eds) *Equity Issues in Radioactive Waste Management*, 301–31. Cambridge, Mass: Oelgeschlager, Gunn and Hain.

Katz, D. (1960). The functional approach to the study of attitudes. *Public Opinion Ouarterly*, 24, 163–204.

Kelly, J.E. (1980). Testimony on behalf of the State of Wisconsin regarding the Statement of Position of the United States Department of Energy in the matter of the proposed rulemaking on the storage and disposal of nuclear wastes. *Docket No. PR50–51*. Washington, DC: US Nuclear Regulatory Commission.

Kemeny, J.G. (1979). *Report of the President's Commission on the Accident at Three Mile Island*. Washington, DC: Government Printing Office.

Kraybill, D.B. (1979). *Three Mile Island: local residents speak out*. Unpublished reports. Social Science Center, Elizabethtown College, Elizabethtown, Penn.

Kunreuther, H., Ginsberg, R., Miller, L., Sagi, P., Slovic, P., Borkan, B. and Katz, N. (1978). *Disaster Insurance Protection: public policy lessons*. New York: Wiley.

Lagadec, P. (1982). *Major Technological Disasters*. Oxford: Pergamon Press.

Langer, E. (1975). *The Illusion of Control*. Beverly Hills, Calif: Sage.

Lathrop, J. and Linnerooth, J. (1983). The role of risk assessment in a political decision process. In Humphreys, P., Svenson, O. and Vari, A. (eds) *Analys-*

ing and Aiding Decision Processes, 39–68. Amsterdam: North-Holland.
Lazarus, R.S. (1966). *Psychological Stress and the Coping Process*. New York: McGraw-Hill.
Lazarus, R.S. and Cohen, J. (1977). Environmental stress. In Wohlwill, J. and Altman, I. (eds) *Human Behavior and Environment* (90–127). New York: Plenum.
Leopold, R.L. and Dillon, H. (1963). Psycho-anatomy of a disaster: a long term study of post-traumatic neurosis in survivors of a marine explosion. *American Journal of Psychiatry*, 119, 913–921.
Leventhal, H. (1970). Findings and theory in the study of fear communications. In Berkowitz, L. (ed.) *Advances in Experimental Social Psychology*, 5, 119–86. New York: Academic Press.
Levine, A.G. (1982). *Love Canal: science, politics, and people*. Lexington, Mass: D.C. Heath.
Lichtenstein, S., Slovic, P., Fischhoff, B., Layman, M. and Combs, B. (1978). Judged frequency of lethal events. *Journal of Experimental Psychology: human learning and memory*, 4, 551–78.
Lifton, R.J. (1976). Nuclear energy and the wisdom of the body. *Bulletin of the Atomic Scientists*, 32, 16–20.
Lifton, R.J. (1979). *The Broken Connection: on death and the continuity of life*. New York: Simon and Schuster.
Lindell, M.K. and Earle, T.C. (1980). *Public Attitudes toward Risk Tradeoffs in Energy Policy Choices*. Batelle Human Affairs Research Centers Report PNL-3402, BHARC-411/80/025.
Lindell, M.K. and Earle, T.C. (1983). How close is close enough: public perceptions of the risk of industrial facilities. *Risk Analysis*, 3, 245–53.
Lowrance, W.W. (1976). *Of Acceptable Risk*. Los Altos, Calif: Kaufman.
Lyons, W., Freeman, P. and Fitzgerald, M.R. (1986). *Public Opinion and the Legislative Response to the Hazardous Waste Challenge*. Paper presented at the American Political Science Association, 26–31 August, Washington DC.
McGrath, J. (1970). *Social and psychological factors in Stress*. New York: Holt.
McGuire, W.J. (1969). The nature of attitudes and attitude change. In Lindzey, G. and Aronson, E. (eds) *The Handbook of Social Psychology*, 2nd edn, 3, 136–314. Reading, Mass: Addison-Wesley.
McKay, J. (1984). Community response to hazard information. *Disasters*, 8(2), 118–23.
Maderthaner, R., Guttman, G., Otway, H. (1978). Effect of distance upon risk perception. *Journal of Applied Psychology*, 3, 380–90.
March, C. and Fraser, C. (eds) (1989). *Public Opinion and Nuclear Weapons*. London: Macmillan.
Marks, G. and Miller, N. (1987). Ten years of research on the false consensus effect: an empirical and theoretical review. *Psychological Bulletin*, 102, 72–90.
Mason, J.W. (1975). A historical view of the stress field, Parts 1 and 2. *Journal of Human Stress*, 1, 6–12, 22–36.

Mazur, A. (1981). *The Dynamics of Technical Controversy*. Washington, DC: Communications Press, Inc.

Mazur, A. (1984). Media influences on public attitudes toward nuclear power. In Freudenberg, W.R. and Rosa, E.A. (eds) *Public Reactions to Nuclear Power: are there critical masses*, 97–114. Boulder, Colo: Westview Press.

Mazur, A. and Conant, B. (1978). Controversy over a local nuclear waste repository. *Social Studies of Science*, 8, 235–43.

Melber, B.D., Nealey, S.M., Hammersla, J. and Rankin, W.L. (1977). *Nuclear Power and the Public: analysis of collected survey research*. PNL-2430 Seattle, Wash: Battelle Human Affairs Research Centers.

Midden, C.J.H. and Verplanken, B. (1986). *Na Tsjernobyl... Enige conclusies over effecten van het ongeluk in Tsjernobyl op de publieke opinie over kernenergie*. Petten: ECN Report ESC-WR-86-23.

Midden, C.J.H. and Verplanken, B. (1990). The stability of nuclear attitudes after Chernobyl. *Journal of Environmental Psychology*, 10, 111–20.

Mishan, E.J. (1976). *Cost–Benefit Analysis*. New York: Praeger.

Mitchell, R.C. (1984). Rationality and irrationality in the public's perception of nuclear power. In Freudenberg, W.R. and Rosa, E.A. (eds) *Public Reactions to Nuclear Power: are there critical masses?* 137–79. Boulder, Colo: Westview Press.

Morell, D. and Magorian, C. (eds) (1982). *Siting Hazardous Waste Facilities: local opposition and the myth of preemption*. Cambridge, Mass: Ballinger.

NEA (1987). *Chernobyl and the Safety of Nuclear Reactors in OECD Countries: report on the NEA group of experts*. Paris: Report of the Nuclear Energy Agency, Organisation for Economic Cooperation and Development.

Nealey, S.M., Melber, B.D. and Rankin, W.L. (1983). *Public Opinion and Nuclear Energy*. Lexington, Mass: Lexington.

Nelkin, D. (ed.) (1979). *Controversy: politics of technical decisions*. Beverly Hills, Calif: Sage.

Nelkin, D. and Pollak, M. (1977). The politics of participation and the nuclear debate in Sweden, the Netherlands and Austria. *Public Policy*, 25, 333–57.

Nelkin, D. and Pollak, M. (1980). Problems and procedures in the regulation of technological risk. In Schwing, R.C. and Albers, Jr., W.A. (eds) *Societal Risk Assessment: how safe is safe enough?*, 233–48. New York: Plenum.

Newsweek (1986). US Fears and Doubts: a newsweek poll. *Newsweek*, 12 May, p. 30.

NRPB (1986). *A Compilation of Early Papers by Members of NRPB Staff about the Reactor Accident at Chernobyl on 26th April 1986*. Report NRPB-M139. Chilton, UK.

Nucleonic Week (1986). Antinuclear fallout from Chernobyl continues to wash over Europe. *Nucleonic Week*, May 1986, 11–13.

NYT (1973). Nader asserts nuclear industry forces public to accept plants. *New York Times* (15 August 1973) 34:4.

Okrent, D. (1987). The safety goals of the US Nuclear Regulatory Commission. *Science*, 236, 296–300.

Oliver, M. (1988). Reported in Chernobyl: the government's response, minutes of evidence, 2, 227. Report of the House of Commons Select Committee on Energy. London: HM Stationery Office.

Otway, H.J. and Cohen, J.J. (1975). *Revealed Preferences: comments on the starr risk-benefit relationship*. International Institute for Applied Systems Analysis, IIASA-RM-75, Laxenburg, Austria.

Otway, H.J. and Fishbein, M. (1976). *The Determinants of Attitude Formation: an application to nuclear power* (Research Memorandum RM76–80). International Institute for Applied Systems Analysis, Laxenburg, Austria.

Otway, H., Haastrup, P., Connell, W., Gianitsopoulos, G. and Paruccini, M. (1987). *An Analysis of the Print Media in Europe following the Chernobyl Accident*. Report of the Joint Research Center of the Commission of the European Community. Ispra, Italy (April 1987).

Otway, H., Maurer, D. and Thomas, K. (1978). Nuclear Power: the question of public acceptance. *Futures*, 10, 109–18.

Otway, H.J. and von Winterfeldt, D. (1982). Beyond acceptable risk: on the social acceptability of technologies, *Policy Sciences*, 14, 247–56.

Pahner, P.D. (1975). The psychological displacement of anxiety: an application to nuclear energy. In Durent, D. (ed.) *Risk–Benefit Methodology and Application: some papers presented at the Engineering Foundation Workshop*, 557–580. Asilomar, C.A. Los Angeles: Report ENO-7589.

Pahner, P.D. (1976). *A Psychological Perspective of the Nuclear Energy Controversy*. International Institute for Applied Systems Analysis Report RM-76-67, Laxenburg, Austria.

Parish, R.M. (1976). The scope of benefit–cost analysis. *Journal of the Economic Society of Australia and New Zealand*, 52, 302–14.

Parker, J. (1978). *The Windscale Inquiry*. London: HM Stationery Office.

Pearlin, L. (1982). The social contexts of stress. In Goldberger, L. and Breznitz, S. (eds) *Handbook of stress*, 367–79. New York: Free Press.

Pearlin, L. and Schooler, C. (1978). The structure of coping. *Journal of Health and Social Psychology*, 84, 627–37.

Pennsylvania Department of Health (1979). *Report on TMI Census Statistics Questionnaires*. Harrisburg, Penn.

Perry, R.W., Greene, M.R. and Lindell, M.K. (1980). Enhancing evacuation warning compliance: suggestions for emergency planning. *Disasters*, 4, 433–49.

Peters, H.P., Albrecht, G., Hennen, L. and Stegelmann, H.U. (1987). *Reactions of the German Population to the Chernobyl Accident: results of a survey*. Nuclear Research Centre Jülich, Jül-Spez-400.

Peters, H.P., Albrecht, G., Hennen, L. and Stegelmann, H.U. (1990). Chernobyl and the nuclear power issue in West German public opinion. *Journal of Environmental Psychology*, 10, 121–34.

Peters, W. (1989). HLW: safe disposal is a reality. *Atom*, 197, November 1989, 18–21.

Philips, L.D. (1984). A theory of requisite decision models. *Acta Psychologica*, 56, 29–48.

Portney, K.E. (1983). *Citizen Attitudes toward Hazardous Waste Facility Siting: public opinion in five Massachusetts communities*. Medford, Mass: Tufts University (Center for Citizenship Public Affairs).
Rankin, W.L. and Nealey, S.M. (1978). Attitudes of the public about nuclear wastes. *Nuclear News*, 21, 112–17.
Ratick, S. and Lakshmanan, T.R. (1980). An overview of the strategic environmental assessment system. In Lakshmanan, T.R. and Nijkamp, P. (eds) *Systems and Models for Energy and Environmental Analysis*, 91–126. Aldershot, UK: Gower.
Renn, O. (1982). Nuclear energy and the public: risk perception, attitudes and behaviour. *Proceedings of the Third Conference of the Uranium Institute, 'Uranium and Nuclear Energy'*, 242–58. London: Butterworth.
Renn, O. (1990). Public responses after Chernobyl: effects on attitudes and public policies. *Journal of Environmental Psychology*, 10, 151–68.
Renn, O. and Hohenemser C. (1987). *Public Responses to Chernobyl: lessons for risk management and communication*. Paper presented at the Annual Conference of the Society for Risk Analysis, Washington, DC.
Renn, O. and Swaton, E. (1984). Psychological and sociological approaches to study risk perception. *Environment International*, 10, 557–75.
Richardson, B., Sorensen, J. and Soderstrom, E.J. (1987). Explaining the social and psychological impacts of a nuclear power plant. *Journal of Applied Social Psychology*, 17, 16–36.
Rippetoe, P.A. and Rogers, R.W. (1987). Effects of components of protection-motivation theory on adaptive and maladaptive coping with a health threat. *Journal of Personality and Social Psychology*, 52, 596–604.
Rogers, G.D. (1984). Residential proximity, perceived and acceptable risk. In Waller, R.A. and Covello, V.T. (eds) *Low Probability/High Consequence Risk Analysis: issues, methods and case studies*, 507–20. New York: Plenum.
Roisser, M. (1983). The use and abuses of polls: a social psychological view. *British Psychological Society Bulletin*, 36, 159–61.
Rosenberg, M.J. (1956). Cognitive structure and attitudinal affect. *Journal of Abnormal and Social Psychology*, 53, 367–72.
Rosenberg, M.J. and Hovland, C.I. (1960). Cognitive, affective and behavioral components of attitudes. In Rosenberg, M.J., Hovland, C.I., McGuire, W.J., Abelson R.P. and Brehm, J.W. (eds) *Attitude Organization and Change: an analysis of consistency among attitude components*, 1–14. New Haven, Conn: Yale University Press.
Roser, T. (1987). The social and political impact of Chernobyl in the Federal Republic of Germany. In Uranium Institute, *Uranium and Nuclear Energy*, Proceedings of the 12th International Symposium. London: Butterworth.
Ross, L. (1977). The intuitive psychologist and his shortcomings: distortions in the attribution process. In Berkowitz, L. (ed.) *Advances in Experimental Social Psychology*, 10. New York: Academic Press.
Ross, L., Greene, D. and House, P. (1977). The 'false consensus effect': an egocentric bias in social perception and attribution processes. *Journal of*

Experimental Social Psychology, 13, 279–301.
Royal Commission on Environmental Pollution (1976). *Sixth Report: nuclear power and the environment*. London: HM Stationery Office.
Russell, M. and Gruber, M. (1987). Risk assessment in environmental policy-making. *Science*, 236, 286–90.
Ryan, A.S. (1984). *Approaches to hazardous waste facility siting in the United States*. Report to the Massachusetts Hazard Waste Facility Site Safety Policy, Boston Mass.
Sandman, P.M. and Paden, M. (1979). At Three Mile Island. *Columbia Journalism Review*, 18, 43–58.
Sassin, W., Hölzl, A., Rogner, H.H. and Schrattenholzer, L. (1983). *Fueling Europe in the Future: the long-term energy problem in EC countries – alternative R and D strategies*. Laxenburg, Austria: International Institute for Applied Systems Analysis, RR-83-9.
Scranton, W.F. (ed.) (1980). *Report of the Governor's Commission on Three Mile Island*. Harrisburg, PA.
Seligman, M.E.P. (1975). *Helplessness*. San Francisco, Calif: Freeman.
Selye, H. (1956). *The Stress of Life*. New York: McGraw-Hill.
Selye, H. (1975). Confusion and controversy in the stress field. *Journal of Human Stress*, 1, 37–44.
Slovic, P. (1987). Perception of risk. *Science*, 236, 280–5.
Slovic, P., Fischhoff, B. and Lichtenstein, S. (1979). Rating the risk. *Environment*, 21, 14–39.
Slovic, P., Fischhoff, B. and Lichtenstein, S. (1980). Facts and fears: understanding perceived risk. In Schwing, R. and Albers Jr., W.A. (eds) *Societal Risk Assessment: how safe is safe enough?*. New York: Plenum Press.
Slovic, P., Fischhoff, B. and Lichtenstein, S. (1982). Facts and Fears: understanding perceived risk. In Kahneman, D., Slovic, P. and Tversky, A. (eds) *Judgment under Uncertainty: heuristics and biases*, 464–89. Cambridge: Cambridge University Press.
Slovic, P., Lichtenstein, S. and Fischhoff, B. (1979). Images of disaster: perception and acceptance of risk for nuclear power. In Goodman, G. and Rowe, W. (eds) *Energy Risk Management*. London: Academic Press.
Soderstrom, E.J., Sorensen, J.H., Copenhaver, E.D. and Carnes, S.A. (1984). Risk perception in an interest group context: an examination of the TMI restart Issue. *Risk Analysis*, 4, 231–44.
Sorensen, J.H. (1983). Knowing how to behave under the threat of a disaster. *Environment and Behavior*, 15, 438–57.
Sorensen, J.H. and Mileti, D.S. (1990). Risk communicating for emergencies. In Kasperson, R. and Stallen, P.J. (eds) *Communicating Risks to the public: International Perspectives*, 367–92. Boston, Mass: Reidel.
Starr, C. (1969). Social benefits versus technological risk, *Science*, 165, 1232–8.
Starr, C. (1980). Risk criteria for nuclear power plants: a pragmatic proposal. *Nuclear Society Conference Paper*. Washington, DC: Nuclear Society.
Starr, C. and Whipple, C. (1980). Risks of risk decisions. *Science*, 208, 1114–9.
Suhonen, P. and Virtanen, H. (1987). *Public Reaction to the Chernobyl Nuclear*

Accident. Tampere, Finland: Research Institute for Social Sciences, University of Tampere.
Sundstrom, E., Lounsbury, J.W., DeVault, R.C. and Peelle, E. (1981). Acceptance of a nuclear power plant: applications of the expectancy value model. In Baum, A. and Singer, J.E. (eds) *Advances in Environmental Psychology*, 3, 171–89. Hillsdale, NJ: Erlbaum.
Sundstrom, E., Lounsbury, J.W., Schuller, C.R., Fowler, J.R. and Mattingly, T.J., Jr. (1977). Community attitudes towards a proposed nuclear power generating facility as a function of expected outcomes. *Journal of Community Psychology*, 5, 199–208.
Swartzman, D., Croke, K. and Swibel, S. (1985). Reducing aversion to living near hazardous waste facilities through compensation and risk reduction. *Journal of Environmental Management*, 20, 43–50.
Tajfel, H. and Wilkes, A.L. (1963). Classification and quantitative judgement. *British Journal of Psychology*, 54, 101–14.
Tasker, A. (1989a). On Nirex safety research. *Atom*, 388 (February) 36–7.
Tasker, A. (1989b). Societal risk and major disasters. *Atom*, 396, October 1989, 10–13.
Thomas, K. and Baillie, A. (1982). Public attitudes to the risks, costs, and benefits of nuclear power. Paper presented at a joint SERÇ/SSRC seminar on research into nuclear power development policies in Britain, June, 1982.
Titchener, J. and Kapp, F.I. (1976). Family and character change at Buffalo Creek. *American Journal of Psychiatry*, 133, 295–9.
Turner, R.H., Nigg, J.M., Pazz, D.H. and Young, B.S. (1981). *Community Responses to Earthquake Threat in Southern California*. Institute for Social Science Research, University of California at Los Angeles.
USDOE (1987). *Health and Environmental Consequences of the Chernobyl Nuclear Power Plant Accident*. US Department of Energy DOE/ER-0332 National Technical Information Service, Springfield, Va.
USEPA (1979). *Draft Environmental Impact Statement for Subtitle C*. Washington, DC: US Environmental Protection Agency.
US National Governors' Association (1981). Siting hazardous waste facilities. *The Environmental Professional*, 3, 133–42.
USNRC (1975). *Reactor Safety Study: an assessment of accident risk in US commercial power plants*. WASH 1400 (NUREQ-75/014). Washington, DC: US Nuclear Regulatory Commission.
USNRC (1986). Policy statement of safety goals. *Federal Register*, 51, p. 28044 (4 August 1986), p. 30028 (21 August 1981). Washington, DC: US Nuclear Regulatory Commission.
USSR State Committee on the Utilization of Atomic Energy (1986). *The Accident at the Chernobyl Nuclear Power Plant and its Consequences*. Report compiled for the IAEA Experts' meeting 25–29 August 1986. Vienna: International Atomic Energy Agency.
Van der Pligt, J. (1985). Public attitudes to nuclear energy: salience and anxiety. *Journal of Environmental Psychology*, 5, 87–97.
Van der Pligt, J. (1988). Applied decision research and environmental policy.

Acta Psychologica, 68, 293–311.

Van der Pligt, J. (1989). Nuclear waste: public perception and siting policy. In Vlek, C.A.J. and Cvetkovich, G. (eds) *Social decision methodology for technological projects*, 235–52. Amsterdam: North-Holland.

Van der Pligt, J. and de Boer, J. (1990). Contaminated soil and the public: risk-perception and policy implications. In Kasperson, R.E. and Stallen, P.J. (eds) *Communicating Risks to the Public*, 127–44. Dordrecht, Netherlands: Kluwer.

Van der Pligt, J. and Eiser, J.R. (1984). Dimensional salience, judgment and attitudes. In Eiser, J.R. (ed.) *Attitudinal Judgment*, 161–78. New York: Springer Verlag.

Van der Pligt, J., Eiser, J.R. and Spears, R. (1986a). Construction of a nuclear power station in one's locality: attitudes and salience. *Basic and Applied Social Psychology*, 7, 1–15.

Van der Pligt, J., Eiser, J.R. and Spears, R. (1986b). Attitudes toward nuclear energy: familiarity and salience, *Environment and Behaviour*, 18, 75–93.

Van der Pligt, J., Eiser, J.R. and Spears, R. (1987a). Comparative judgments and preferences: the influence of the number of response alternatives. *British Journal of Social Psychology*, 26, 269–80.

Van der Pligt, J., Eiser, J.R. and Spears, R. (1987b). Nuclear waste: facts, fears and attitudes. *Journal of Applied Social Psychology*, 17, 453–70.

Van der Pligt, J., Ester, P. and van der Linden, J. (1983). Attitude extremity, consensus and diagnosticity. *European Journal of Social Psychology*, 13, 437–9.

Van der Pligt, J., van der Linden, J. and Ester, P. (1982). Attitudes to nuclear energy: beliefs, values and false consensus. *Journal of Environmental Psychology*, 2, 221–31.

Van Schie, E.C.M. and van der Pligt, J. (1992). Getting an anchor on availability. *Organizational Behavior and Human Decision Processes*, in press.

Verplanken, B. (1989). Beliefs, attitudes, and intentions toward nuclear energy before and after Chernobyl in a longitudinal within-subjects design. *Environment and Behavior*, 21, 371–92.

Vlek, C.A.J. (1986). Risk, decline and aftermath of the Dutch 'Societal Discussion' on (nuclear) energy policy. In Becker, H.A. and Porter, A. (eds) *Impact Assessment Today*, 141–88. Utrecht, Netherlands: Van Arkel.

Vlek, C.A.J. and Otten, W. (1987). Judgmental handling of energy scenarios: a psychological analysis and experiment. In Wright, G. and Ayton, P. (eds) *Judgmental Forecasting*, 267–89. Chichester, UK: Wiley.

Vlek, C.A.J. and Stallen, P.J. (1981). Judging risks and benefits in the small and large, *Organizational behavior and Human Performance*, 28, 235–71.

Von Winterfeldt, D. and Edwards, W. (1984). Patterns of conflict about risky technologies. *Risk Analysis*, 4, 55–68.

Von Winterfeldt, D. and Edwards, W. (1986). *Decision Analysis and Behavioral Research*. Cambridge: Cambridge University Press.

Von Winterfeldt, D., John, R.S. and Borcherding, K. (1981). Cognitive components of risk ratings. *Risk Analysis*, 1, 227–87.

Wallmann, W. (1987). Chernobyl's impact in West Germany. *Forum for Applied Research in Public Policy*, Summer 1987, 13–15.
Warren, D.S. (1981). Local attitudes to the proposed Sizewell 'B' nuclear reactor. *Report RE19*, Food and Energy Research Centre, October 1981.
Weinberg, A. (1972). Social institutions and nuclear energy. *Science*, 177, 32–3.
Weinberg, A. (1977). Is nuclear energy acceptable? *Bulletin of the Atomic Science*, 33, 54–60.
Weinstein, N.D. (1980). Unrealistic optimism about future life events. *Journal of Personality and Social Psychology*, 39, 106–20.
Welch, M.J. (1985). Nuclear waste disposal reaches critical stage. *New York Times* (20 March), section A, 26.
WHO (1986). *Chernobyl Reactor Accident: reports of a consultation*, Report ICP/CEH 139. World Health Organization, Regional Office for Europe, Copenhagen.
Wilson, R. (1986). Chernobyl: assessing the accident. *Issues in Science and Technology*, 3, 21–9.
Witkin, H.A., Goodenough, D.R. and Oltman, D.K. (1979). Psychological differentiation: current status. *Journal of Personality and Social Psychology*, 37, 1127–45.
Woo, T.O. and Castore, C.H. (1980). Expectancy-value and selective exposure determinants of attitudes toward a nuclear power plant. *Journal of Applied Social Psychology*, 10, 224–34.
Wynne, B. (1982). *Rationality and Ritual: the Windscale Inquiry and nuclear decisions in Britain*. Chalfont St Giles, Bucks: The British Society for the History of Science.
Wynne, B. (1989). Sheepfarming after Chernobyl. *Environment*, 31, no. 2, 11–39.

Index of subjects

affect, 103–4
attitudes, 39–41
 and accidents, 7–9, 33–4, 57–8, 119–20, 126
 and equity, 82–5
 to local power plants, 3–4, 17, 59–61, 63–5
 to nuclear energy, 2–7, 41–3
 and risks, 6, 45, 82–5
 and salience, 43–4
 structure of, 42–51, 65–8
 and values, 44–51
 to waste facilities, 9–11, 79–82
attributions, 56
availability, 16, 26–7

biases, in decision-making, 51–3
 in risk perception, 23–7
 in social perception, 53–6

CEGB, 59, 69–70
chemical plants, 60
chemical waste, 31–2, 85, 89, 107
Chernobyl, accident at, 7–9, 106, 120–5
 and public opinion, 7–9, 124–6
 and risk communication, 126–30, 134–8
coping styles, 101, 103
 and stress, 102–5

decision aids, 154–66
decision analysis,
 and cost-benefit analysis, 162–4
 and MAUT, 160–2
 and risk assessment, 159–60
 and scenarios, 154–8
 and values, 165–6
defensive avoidance, 101
disaster warnings, 143–5
dissonance, 43, 69, 87
dose-response relations, 18–19

economic compensation, 89–90
emergency information, 140–5
environmental stressors, 95–8
equity, 37, 82–9
estimates of risks,
 experts' estimates, 25, 27, 30–1, 36
 lay estimates, 25, 27, 30–1
 technical estimates, 25, 159–60
expectancy value, 40–1
expressed preferences, 25–6

false consensus, 54–7
familiarity,
 and attitudes, 68–74
 and dissonance, 69
 and public opinion, 11
 and risk perception, 27–8, 31–3

Index of subjects

and salience, 68–74
frame of reference, 35, 139–40
framing, 16

groupthink, 52–3

heuristics, 23, 156
hypervigilance, 101

information processing, 43–4,
  53–4, 103
irrational fear, 11–14, 84

learned helplessness, 101–2, 118
local attitudes,
  during construction, 61–3
  during siting, 63–5
Love Canal, 107, 113, 143

'NIMBY'-effect, 4, 60, 76
nuclear waste management, 9–11
  and equity, 82–5
  and public opinion, 9–11, 79–83
  and siting, 79–93

perceived control, 28–30, 97–8,
  101–2, 107, 117–19
perseverance of attitudes, 51–3
policy decision-making, 145–50,
  159–60
  and conflicts, 34–7
  and risk assessment, 159–60
probability assessment, 157–8
public inquiries, 1, 93, 146
public opinion to nuclear energy,
  2–7
  explanations of, 11–14, 84
  and familiarity, 11, 68–74
  and health and safety, 14, 62–5,
    67–8, 72–4
  and nuclear accidents, 7–9,
    119–20, 124–6
  to nuclear power stations, 4–5,
    61–3, 65–8, 74–5
  to nuclear waste, 9–11, 79–82
public participation, 92–3, 146–50

quality-adjusted life-years, 22

reactor safety study, 20
risks,
  acceptability of, 6–7, 24–34
  analysis of, 18–23, 34–7
  assessment, 18–23, 159–60
  and benefits, 41–3, 85–91
  communication of, 91–2, 132–44
  coping with, 101, 103
  definitions of, 21–3, 36
  dimensions of, 24–34
  and equity, 37, 82–91
  of opinion research, 15
  perception of, 14–15, 23–34

salience of beliefs, 37, 43–51, 58
  dimensional, 43–4
  and local attitudes, 65–8
selective exposure, 43
Sellafield, 33, 74, 76, 107, 109, 122,
  137–8, 146
siting procedures, 89–93
stress, definitions of, 95
  and coping 100–1
  effects of, 102–5, 115–19
  and nuclear accidents, 106–9,
    115–19, 125–6
  and nuclear energy, 105–10
  and policy decision-making, 110
  theoretical perspectives on,
    98–102
subjective expected utility, 40, 46

technological risks, 24–31
Three Mile Island, consequences of,
  114–15, 119–20
  and public opinion, 2–5, 106
  and risk communication, 132–4
  and stress, 106

unrealistic optimism, 141

values, 12, 35
  and risk analysis, 34–7
  role of, 12, 34–7, 49–51
VOSDA, 165–6

Index of names

Adler, A., 112
AIF, 21
Aikin, A.M., 76
Ajzen, I., 40–1
Allen, F., 152
Allport, G.W., 40
Appley, M., 95
Atom, 3, 11
Averill, J.R., 101

Bachrach, K.M., 90
Baillie, A., 11, 16, 69
Barnes, K., 132, 134, 141, 143–4
Baum, A., 96–7, 99, 102–3, 106, 112–13, 116–18
Baum, C., 96–7
Baxter, R.K., 84, 88
Bertell, R., 34
Bishop, W.P., 82
Black, J.S., 91
Borcherding, K., 141
Bord, R.J., 83, 89–91
Bowonder, B., 133, 141
Braine, B., 86, 88
Broadbent, D., 103
Bromet, E., 116
Bromet, E.J., 87
Brooks, H., 82–3
Brown, J., 84
Burns, M.E., 89

Campbell, J., 96–7, 102
Carle, R., 90
Carnes, R., 91
Castore, C.H., 51, 63–4, 68
Checkoway, B., 92
Chen, K., 165
Chess, C., 151
Cohen, B.L., 83
Cohen, J.J., 24
Cohen, S., 95–105
Collins, D.L., 116–18
Colson, J.P., 146
Combs, B., 26
Commission of the European Communities, 5, 6, 53
Conant, B., 82–3
Covello, V.T., 133, 138, 141–3, 151–2
Croke, K., 90
Cunningham, S., 84
Cutter, S., 132, 134, 141, 143–4
Cvetkovich, G., 142, 152

Daamen, D.D.L., 27
Davidson, L.M., 112–13, 116–18
Dawes, R.M., 39
De Boer, J., 107, 113
Decima, 3, 5, 9–10, 14, 20, 53, 79
Derr, P., 86
Dew, M.A., 87

Dillon, H., 113
Disabella, R., 114–15
DOE, 123
Dohrenwend, B.P., 116, 119–20
Douglas, M., 35
Drottz, B.M., 125
Ducot, C., 156
Dupont, R.L., 12, 83–4

Earle, T.C., 31–2, 51, 74, 87, 142, 152
Edwards, W., 23, 34–6, 38, 40, 153, 161
Eiser, J.R., 11, 16–17, 35, 43–5, 49, 53–5, 59–60, 65–73, 88, 96, 106, 141
Emmings, A., 81
Enbar, M., 86
Ester, P., 35, 46–7, 49–50, 55–7, 69, 87
Evans, G.W., 95–105
Eyre, B., 82

Fallows, S., 79–80
Fessenden-Raden, J., 89
Festinger, L., 43, 69
Fielding, J., 84
Fischhoff, B., 26–30, 33, 36–7, 92, 141, 157, 161
Fishbein, M., 40–2
Fisk, S., 43
Fitchen, J., 89
Fitzgerald, M.R., 87, 90
Flavin, C., 127
Fleming, R., 99, 112–13, 117–18
Flowers, R., 82
Flynn, C.B., 115–16, 119, 141, 144
Foa, F.B., 90
Foa, U.G., 90
Folkman, S., 97
Fowlkes, M., 90
Fraser, C., 12
Freedman, J.L., 43
Freeman, P., 87–90
Freudenburg, W.R., 84, 88

Gardner, M.J., 36
Gatchel, R.J., 106, 117
Glass, D., 99–100, 104, 118
Gleser, G., 113
Goldhaber, M.K., 114–15, 117
Goodenough, D.R., 117
Granberg, D., 13
Green, B.L., 113
Greene, D., 55–6
Greene, M.R., 144
Groth, A.J., 83
Gruber, M., 19, 20, 160
Grundberg, N., 102–3
Guttman, G., 142

Halmberg, S., 13
Hance, B.J., 151
Hare, F.K. 76
Harnden, D.G., 35
Harris and Associates, 83
Heath, J., 89
Henderson, J., 84
HMSO, 160
Hodler, T.W., 144
Hohenemser, C., 12, 31, 123–4, 127–9
House, P., 55–6
Houts, P.S., 114–17
Hovland, C.I., 40
Hu, T., 115
Hughey, J.B., 11, 61–3, 69
Hunter, 40

Jacobs, S.V., 101
Jacobson, N., 34
Janis, I.L., 52, 101, 103, 161
Jepson, C., 158
John, R.S., 141
Johnson, B.B., 84, 88–9, 91
Jungermann, H., 156, 158

Kahneman, D., 23, 157
Kaplan, H.B., 95
Kapp, F.I., 11

Index of names

Kasperson, R.E., 1, 9, 12, 81–2, 85–6, 90–2, 149
Kates, R.W., 12, 31, 86, 88
Katz, D., 40
Kelly, J.E., 91
Kemeny, J.G., 114–15
Krantz, D.H., 158
Kraybill, D.B., 116
Kunreuther, H., 141

Lagadec, P., 133
Lakshmanan, T.R., 155
Langer, E., 101
Lathrop, J., 159
Lazarus, R.S., 95–7, 99, 104, 118
Lee, I.S., 83
Leopold, R.L., 113
Leventhal, H., 101
Levine, A.G., 107, 113, 143
Lichtenstein, S., 25, 26–30, 33, 37, 92, 141, 157, 161
Lifton, R.J., 12, 29
Lindell, M.K., 31–2, 51, 74, 87, 144
Linnerooth, J., 159
Lounsbury, J.W., 61–3
Lowrance, W.W., 24, 27
Lubben, H.J., 156
Lyons, W., 87, 90

McGrath, J., 104
MacGregor, D., 27
McGuire, W.J., 43
McKay, J., 142
Maderthaner, R., 142
Magorian, C., 87, 89
Mann, L., 101, 103, 161
March, C., 12
Marks, G., 56
Mason, J.W., 99
Mathes, J.C., 165
Maurer, D., 41–2, 51, 66
Mazur, A., 35, 82–3, 120
Melber, B.D., 2–3, 9–12, 14, 17, 53, 83
Midden, C.J.H., 15, 17, 27

Mileti, D.S., 133, 142–3
Miller, N., 56
Miller, P., 90
Mishan, E.J., 164
Mitchell, R.C., 13, 84
Montgomery, H., 156
Morell, D., 87, 89

NEA, 123
Nealey, S.M., 2, 3, 9–12, 14, 17, 53, 83
Nelkin, D., 35, 146–8, 150
Newsweek, 9
Nisbett, R.E., 158
NRPB, 136
Nucleonic Week, 9
NYT, 80

Okrent, D., 20
Oliver, M., 137
Oltman, D.K., 117
Otten, W., 156
Otway, H.J., 24, 35, 41–2, 51, 66, 125–7, 142

Paden, M., 144
Pahner, P.D., 29, 83
Parish, R.M., 164
Parker, J., 44
Pearlin, L., 95, 97
Perry, R.W., 144
Peters, H.P., 125–6, 128–9
Peters, W., 156
Philips, L.D., 166
Pollak, M., 146–8
Ponzurick, P.J., 83, 89
Portney, K.E., 84, 91

Rankin, W.L., 2, 3, 12, 14, 17, 53, 83
Ratick, S., 155
Renn, O., 8, 9, 21, 36, 101, 121, 124–30, 141
Richardson, B., 165
Rippetoe, P.A., 101
Rogers, G.D., 87

Index of names

Rogers, R.W., 101
Roisser, M., 16
Rosenberg, M.S., 40, 46
Roser, T., 128
Ross, L., 51, 55–6
Royal Commission on Environmental Pollution, 82
Russell, M., 19, 20, 160
Ryan, A.S., 81

Sandman, P., 144, 151
Sassin, W., 155
Schaeffer, M.A., 106, 117
Schooler, C., 97
Schutz, H.G., 83
Scranton, W.F., 115
Sears, D., 43
Seligman, M.E.P., 101–2
Selye, H., 98–9
Singer, J.E., 96–7, 99–100, 102–4, 117–18
Sjöberg, L., 125
Slovic, P., 18, 23, 26–7, 30, 33, 37, 92, 133, 138, 141–3, 151, 157, 161
Smith, T.L., 39
Soderstrom, E.J., 88, 165
Sorensen, J.H., 133, 142–3, 165
Spacapan, S., 104
Spears, R., 11, 16–17, 59–60, 65–73, 88, 96, 106, 141
Stallen, P.J., 30, 33
Starr, C., 24, 27, 90
Stogre, M., 34
Suhonen, P., 8
Sundstrom, E., 51, 61–3
Swartzman, D., 90
Swaton, E., 21
Swibel, S., 90

Tajfel, H., 43
Tasker, A., 21, 22, 76

Taylor, S.E., 43
Thomas, K., 11, 16, 41–2, 51, 66, 69
Titchener, J., 113
Trumbull, R., 95
Turner, R.H., 141
Tversky, A., 23, 157

US National Governors' Association, 85
USEPA, 81
USNRC, 20, 21

Van der Linden, J., 35, 46–7, 49–50, 55, 57
Van der Pligt, J., 11, 16–17, 43–50, 53–7, 59–60, 65–73, 88, 92, 96, 106–7, 113, 141, 157–8
Van Schie, E.C.M., 157
Verplanken, B., 15, 17, 27
Virtanen, H., 8
Vlek, C.A.J., 30, 33, 92–3, 156
Von Winterfeldt, D., 34–6, 38, 133, 138, 141–3, 153, 161

Wallmann, W., 128
Warren, D.S., 11, 69
Weinberg, A., 78, 82
Weinstein, N., 141
Weinstein, N.D., 101
Welch, M.J., 79
Wildavsky, A., 35
Wilkes, A.L., 43
Wilson, R., 123
Winget, C.W., 113
Witkin, H.A., 117
Witzig, W.F., 83, 89
Woo, T.O., 51, 63–4, 68
Wynne, B., 93, 123, 129, 134–8

Zautra, A.J., 90